Mona Numairy
Mo'awia Hassan

Genetic diversity of Phlebotomus orientalis in Sudan

AF138174

Mona Numairy
Mo'awia Hassan

Genetic diversity of Phlebotomus orientalis in Sudan

Population Structure of Kala-azar vector in Sudan

LAP LAMBERT Academic Publishing

Impressum / Imprint

Bibliografische Information der Deutschen Nationalbibliothek: Die Deutsche Nationalbibliothek verzeichnet diese Publikation in der Deutschen Nationalbibliografie; detaillierte bibliografische Daten sind im Internet über http://dnb.d-nb.de abrufbar.

Alle in diesem Buch genannten Marken und Produktnamen unterliegen warenzeichen-, marken- oder patentrechtlichem Schutz bzw. sind Warenzeichen oder eingetragene Warenzeichen der jeweiligen Inhaber. Die Wiedergabe von Marken, Produktnamen, Gebrauchsnamen, Handelsnamen, Warenbezeichnungen u.s.w. in diesem Werk berechtigt auch ohne besondere Kennzeichnung nicht zu der Annahme, dass solche Namen im Sinne der Warenzeichen- und Markenschutzgesetzgebung als frei zu betrachten wären und daher von jedermann benutzt werden dürften.

Bibliographic information published by the Deutsche Nationalbibliothek: The Deutsche Nationalbibliothek lists this publication in the Deutsche Nationalbibliografie; detailed bibliographic data are available in the Internet at http://dnb.d-nb.de.

Any brand names and product names mentioned in this book are subject to trademark, brand or patent protection and are trademarks or registered trademarks of their respective holders. The use of brand names, product names, common names, trade names, product descriptions etc. even without a particular marking in this work is in no way to be construed to mean that such names may be regarded as unrestricted in respect of trademark and brand protection legislation and could thus be used by anyone.

Coverbild / Cover image: www.ingimage.com

Verlag / Publisher:
LAP LAMBERT Academic Publishing
ist ein Imprint der / is a trademark of
OmniScriptum GmbH & Co. KG
Heinrich Böcking Str. 6 8, 66121 Saarbrücken, Deutschland / Germany
Email: info@lap-publishing.com

Herstellung: siehe letzte Seite /
Printed at: see last page
ISBN: 978-3-659-76184-3

Zugl. / Approved by: Khartoum, University of Khartoum, Ph. D. thesis, 2009

Copyright © 2015 OmniScriptum GmbH & Co. KG
Alle Rechte vorbehalten. / All rights reserved. Saarbrücken 2015

DEDICATION

To my parents for their continuous love, support and blessings

To my brothers and sisters, for their tender care and encouragement

To those who believes in my capabilities

To all the faithful and worm hearts who surrounded me by their real love and care

&

To all whom I love

Mona Siddig

AKNOWLADGEMENT

Most of all thanks are due to *Allah* the most helper and the most merciful

I would like to express my gratitude and sincere appreciation to my supervisor *Dr. Mo'awia Mukhtar Hassan* for his invaluable guidance and continuous support and encouragement, without his help, valuable suggestions and discussion, generous advices through the course of this study, this work would have never been accomplished. Also thanks and appreciation is extended to my co-supervisor *Dr. Omran Fadl Osman* of the University of Khartoum, zoology department. For his valuable advises and help.

I would like to express my deep thank to *Dr. Intisar. E. El Rayah*, the Director of Tropical Medicine Research Institute (TMRI), for her hospitality and co-operation. My thanks are extended to *Dr. Badria Babiker El-Sayed*, the head of Epidemiology department and *Dr. Atif A. Elagib* head of Immunology and Biotechnology in TMRI and their staff for their help and valuable advices.

I would like to express my deep thanks to *Prof. Halla Kasim*, of Institute of Environmental Studies and Research, University of Ain Shams, for her great help.

I am indebted to my colleagues in sandfly research group *Mr. Osman Mohi Eldin*, *Mr. Rida Mohamed Elhag*, *Miss Sally O. Widaa* and finally to *Miss Mihad Abdelaal Ibrahim* for their support in the field and laboratory work.

Special thanks to *Mr. Salah Eldin Gumaa Elzaki*, for his technical support, good discussion and advices. My thanks also extended to *Mr. Tellal Babiker Ajeeb*, *Mr. Samir Abu Elgasim Hammouda* and *Mr. Haitham Elbir* for their great help and support.

I am grateful to *Dr. Sahar Mubarak Bakhet* and *Dr. Hisham Yousif* of the Institute of Endemic Diseases, University of Khartuom for their great help in data analysis.

Thanks are also extended to *Dr. Nawal Tag Elsir* of the National Health Laboratory for her hospitality and invaluable support and encouragement throught the course of this work.

I am grateful to *Miss. Sababul* the laboratory technician of Doka hospital, Gadarif State for her good hospitality during samples collection.

I would like to thanks *Mr. Brema Musa* the technician in the Department of Zoology, University of Khartoum for his great help in sandflies collection.

I am grateful to the community of Sorugia village and Gadarif area, for their continuous help during sample collection.

This study received financial support from **UNICEF/UNDP/World Bank WHO**. Special programme for Research and Training in Tropical Disease (TDR). RCS/re-entry grant (**Mo'awia M. Hassan: A41387-TDR, WHO**).

LIST OF CONTENTS

LIST OF FIGURES

LIST OF TABLES

CHAPTER ONE

INTRODUCTION AND LITRATURE REVIEW

1.1. General introduction:

Leishmaniasis is a globally widespread group of diseases affecting humans and animals. The disease is caused by flagellate protozoa belonging to the genus *Leishmania* (Trypansomatidae; Kinetoplastidae) (WHO, 2006). Leishmaniasis is transmitted by a bite of an infected female phlebotomine sandfly. Rodents and canines are incriminated as reservoir hosts, whereas probable anthroponotic transmission in restricted foci and under certain circumstances may takes place (Desjeux, 2001).

The disease manifests itself in a variety of clinical forms, ranging from a self-healing cutaneous lesions (CL) to a fatal visceralizing form (VL; Kala-zar), and also includes a metastasizing muco-cutaneous form (MCL), and a post kala-azar dermal leishmaniasis (PKDL). Recently, *Leishmania*/HIV co-infection is emerging as a new clinical form of leishmaniasis (WHO, 2006).

The World Health Organization (WHO) considered leishmaniasis to be one of the most important parasitic diseases (WHO, 2002). The leishmaniasis is endemic in 88 countries in the four continents (22 in New World and 66 in Old World) with an estimated yearly incidence of 1-1.5 million cases of CL and 500,000 cases of VL (WHO, 2006). The population at risk is estimated at 350 million people (Desjeux, 1996; WHO, 2006) with an overall prevalence of 12 million cases worldwide (WHO, 2000). Over 90% of the cases of CL occur in Afghanistan, Algeria, Brazil, Iran, Peru, Saudi Arabia and Syria, while 90% of the visceral cases occur in Bangladesh, Brazil, India, Nepal and Sudan (WHO, 2006).

In Sudan, leishmaniasis is endemic in several regions. The visceral leishmaniasis (VL) represents the serious health problem causing high numbers of fatalities. The main endemic focus of VL extends from the Upper Nile State in Southern Sudan east to the Blue Nile State reaching Gedaref State in the North east of Sudan (El-Hassan *et al.*, 2001a). Other small foci include the Nuba Mountain and Darfur in the West (Desjeux, 1991). Large epidemic of VL occurred in Upper Nile in 1990's and claimed the lives of more than 100,000 individuals (Seaman *et al.*, 1996). Reports from Medicine Sans Frontiers- Holland (MSF-Holland) showed that during 1996-1999, 14,000 of VL cases occurred in a population of 200,000 people in eastern Sudan (Elnaiem *et al.*, 2003).

In Sudan, VL is caused by *Leishmania donovani* zymodemes belongs to three phylogenic complexes; *L. donovani* (MON-18; MON-274) MON-276), *L. infantum* MON-30; MON -81; MON-267 and MON-278) and *L. archibaldi* (MON-82; MON-

1

257 and MON-258) (Pratlong *et al.*, 2001; Dereure *et al.*, 2003). The proven vector is *Phlebotomus (Larroussius) orientalis*, which has been shown to thrive in *Acacia seyal-Balanites aegyptiaca* woodland (Hoogstraal & Heyneman, 1969; Elnaiem *et al.*, 1998a, 1998b; Thomson *et al.*, 1999).

Leishmaniasis control is mainly based on treatment of the *Leishmania* infected cases. However, the most successful control methods is relay on combination of treatment of the patients and where feasible with vector control and reservoir hosts (WHO, 2006). In many leishmaniasis endemic areas such as India and Sudan the sandfly control is a collateral benefit of malaria control using different insecticides (WHO, 2006). Moreover, self-protection can be done by using impregnated bednets and curtains with insecticides such as permethrin and deltamethrin (WHO, 2006).

1.2. Vectors of leishmaniasis:

Phlebotomine Sandflies are insects of the order Diptera, family Psychodidae and subfamily Phlebotominae. Sandflies include well-known vectors of pathogens and parasites responsible for human and animal diseases such as leishmaniasis in many regions of the world (Killick-Kendrick, 1990). Sandflies represent vectors belongs to two genera, *Phlebotomus* (Old World) and *Lutzomyia* (New World). There are more sandfly species in the New World than in the Old World and this is often quoted as the reason that there are more species of *Leishmania* species in the New World (León *et al.*, 1996). The most widely accepted classification of phlebotomine is that of Lewis *et al.* (1977) which is based on morphological adult characters and on traditional systematic opinion (i.e. regardless of phylogenetic affiliations among taxa).

Sandfly vectors of visceral leishmaniasis due to *L. donovani* are: *P. argentipes* in India, *P. chinensis* in China, *P. perniciosus* in North Africa, Italy, France and Portugal, *P. perfiliewi* in Greece, *P. orientalis* in Sudan and Ethiopia, *P. martini* in Kenya (Le Blancq & Peters, 1986). *Leshmania infantum* is transmitted by *P. perniciosus*, *P. ariasi*, *P. perfiliewi* and *P. neglectus*, whereas, *L. chagasi* is transmitted by the *L. longipalpis* (WHO, 1990). Sandflies usually repose during the day in burrows, tree hollows, caves or buildings. After sundown, they leave the shelters to remain active throughout the night.

1.2.1. General biology:

Phlebotomine sandflies are members of family Psychodidae, order Diptera. They are delicate insects and can be distinguished from other Nematocerous Dipterans by the brownish colour, small size (1.5-2.5 mm), hairy appearance, long slender legs, jerky flight pattern and the characteristic manner in which they hold their pointed wings at an angle of 45° above their body (like a vertical V) (Lane,

2

1993).

Sandflies are grouped in six genera namely, *Lutzomyia, Brumptoyia* and *Warilya* (nearetic genera); and *Phlebotomus, Sergentomyia* and *chinus* (palearetic genera). Out of 800 species of sandflies exist in the world, only about 70 species of *Phlebotomus* and *Lutzomyia* are known as vectors of diseases of man and animals (Killick-Kendrick, 1990; Lane, 1993).

1.2.2. Life history:

After copulation and blood feeding, females rest for 2-3 days to mature eggs. The female sandfly lays between 30-70 eggs and hatching occurs one to two weeks later, depending on the environmental conditions (Lane, 1993). At 28 °C, the larva hatches from the egg and starts to feed on organic matters (Lewis, 1973). There are four larval instars of sandflies which feed on organic detritus and micro-organisms. The larva develops in 30-40 days into a pupa which hatches into a fully grown adult in one week (Lewis, 1973; Hassan, 2004). The period from oviposition to adult eclosion is 25-60 days (Hassan, 2004), but may extend up to several months in dipausing species (Ready, 1979). The adult emerges from the pupa during the hours of darkness often just before dawn (Lane, 1993). In males, the terminalia rotate through 180° during the first 24 hours after emergence. Most species are gonotrophically concordant, but *P. papatasi* may feed more than once in an ovarian cycle (Figure 1).

1.2.3. Environmental factors and distribution of sandflies:

Phlebotomine sandflies occur through the tropical, subtropical and temperate parts of the world, however, some species penetrate into the temperate regions (50° N and 40° S) (Lane, 1993). The spatial distribution and seasonal dynamics of sandflies are influenced by environmental factors (i.e. wind, temperature, relative humidity, soil texture and vegetation cover) (Wasserberg *et al.*, 2003; Gebre-Michael *et al.*, 2004) and by biological factors such as blood and sugar sources, resting sites and breeding sites. Moreover, urbanization conditions including human activities expansion, habitat fragmentation and deforestation led to the emergence and reemergence of some sandfly vectors in many areas worldwide (Tauil, 2006).

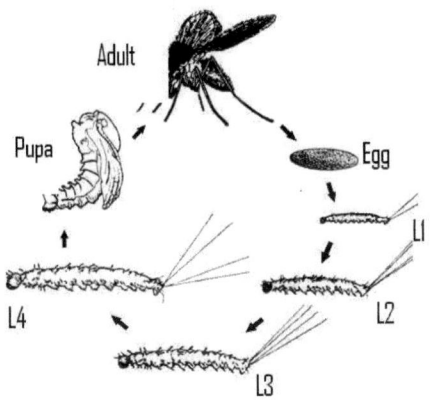

L1= indicates 1ˢᵗ larval instar
L2= indicates 2end larval instar
L3= indicates 3ʳᵈ larval instar
L4= indicates 4th larval instar

Figure 1: Lifecycle of sandflies (www.liv.ac.uk)

Climatic conditions are of the most important factors that determine the distribution of the phlebotomine sandflies. For example the temperature and relative humidity are known to affect the survival of sandflies and the speed of development of different stages in the life cycle (Kirk & Lewis, 1947). The tropical species are known to thrive at a constant temperature (Ward, 1989) whereas the larvae of European species such as *P. ariasi* maintain over wintering diapause (Kellick-Kendrick & Kellick-Kendrick, 1987). Elnaiem *et al.* (1998b) reported that in Sudan, the sites of occurrence of *P. orientalis* characterized by higher maximum and minimum daily temperature than the negative sites.

Also, the soil texture and the associated vegetations are major factors that influence the distribution of sandflies (Elnaiem *et al.*, 1998b; Thomson *et al.*, 1999). The two authors found that the distribution of *P. orientalis* in Sudan is associated with *Acacia seyal/Balanites aegyptiaca* that grow on black cotton soil. Ashford and Bettini (1987) also suggested that such association may be due to an indirect influence of soil types on the local microclimates such as vegetation. The deep cracks appear in the soil during the dry season are known to harbor adult sandflies which depend on some mammals found in such habitats (Quate, 1964) and nutrients in the soil for larval development.

Moreover, these environmental factors may have direct impact on the genetic

4

constitution of insect populations which may change through time and space from one population to another and from generation to generation. Therefore, different populations of a single species will be expected to be different in evolutionary stages of divergence mainly due to the intrinsic genetic adaptation to their slight different micro and extrinsic environment (Benzie & Wakeford, 1997).

1.3. History of species complexes:

Species complexes are common in the class insecta (Subbarao, 1997), where morphologically identical populations do not inter breed, a separate biological species may exit, are known as sibling, cryptic or isomorphic species. The first morphological variation in sandflies was described by Mangabeira (1969) in males *Lu. longipalpis* collected from two geographic regions in Brazil. Mangabeira (1969) reported the existence of one- or two-spot phenotypes in male *Lu. longipalpis*, where the spot represents pair of pale patches on the abdominal tergites . The author suggested that these two forms might represent two different species that could be found in different ecological habitats.

Studies on the distribution of the *Lu. longipalpis* indicated that males with one pair of spots have appeared from Mexico to Southern Brazil and that the two-spotted form is concentrated more in northeast Brazil and the two forms occur sympatrically in some Brazilian States (Ward *et al.*, 1985). The intermediate form occurs sympatrically with the two previous forms in northeast Brazil (Ward *et al.*, 1988; Mukhopadhyay *et al.*, 1998).

Later these spots in *Lu. longipalpis* were found to be associated with pheromone disseminating structures (Ward *et al.*,1991). However, the populations of males with different spot morphology but similar pheromones could interbreed and produce fertile progeny, and the taxonomic role for this morphological feature thus became redundant (Ward *et al.*, 1988).

Lanzaro *et al.* (1993) traced the isoenzymatic profiles of populations of *Lu. longipalpis* reared in the laboratory, originally from Costa Rica, Colombia, and Brazil. Some of 27 enzymes were assayed, and the results showed multiple genetic polymorphisms with hybridization between populations resulting in sterile males. This finding suggested that the Costa Rican population was different from those of Brazil and Colombia. Moreover, Warburg *et al.* (1994) reported differences between these populations in the nucleotide sequence for maxadilan, revealing a polymorphism in the Costa Rican population.

Yin *et al.* (1999) examined microscopically the brain cells of the fourth instar sandfly larvae of *L. longipalpis* from Brazil, Colombia and Costa Rica. The author found differences in the G-banding and/or position of the centromere on chromosome four that distinguished four putative sibling species. Furthermore, Lanzaro *et al.*

(1999) reported variation in the primary DNA and inferred amino acid sequences of maxadilan. Differences were found within and among natural field populations which indicated a high degree of divergence in the salivary peptide maxadilan in populations of the Brazil, Colombia and Costa Rica. Mutebi *et al.* (1999) analyzed eleven populations of *L. longipalpis* from different areas of Brazil. Genotypic frequencies within the populations were in close compliance to Hardy-Weinberg expectations, suggesting the occurrence of non sympatric species among these populations. The levels of genetic distance between pairs of populations were very low, consistent with local populations within a single sand fly species. Estimate of effective migration rates among all populations were low, suggesting that gene flow is restricted among populations, which is probably the reason for the observed genetic sub-structuring.

RAPD has been successfully used by Adamson *et al.* (1993) to separate two closely related species of the New World sandfly species, *Lu. yougi* and *Lu. spinacrassa*. Population studies of the New World sandfly species *Lu. longipalpis* and *Lu. whitmani* are now being done with the help of RAPD-PCR (Dias *et al.,* 1998; Margonoria *et al.,* 1998).

1.4. Sandflies and leishmaniasis in Sudan:
1.4.1. Sandflies of Sudan:

The *Phlebotomus* sandflies in the Sudan are *P. rodhaini, P. orientalis, P. pedifer, P. alexandri, P. saevus, P. papatasi, P. bergeroti, P. duboscqi,* and *P. martini.* Out of these species only two species are thought to play a major role in disease transmission, namely *P. orientalis* as a vector of VL (Hoogstraal & Heyneman, 1969; Hassan *et al.,* 2008) and *P. papatasi* as a vector of CL (El-Sayed, 1991). Other *Phlebotomus* species such as *P. saevus, P. duboscqi* and *P. bergeroti* recorded from the Sudan are known as potential vectors of *Leishmania* parasites in neighboring African countries although they are not vectors in Sudan. However, there were many species other than *P. orientalis* which were suspected at some stage as vectors of VL in Sudan. These species were namely *P. martini, P. heischi, P. lesleyae* and *P. papatasi,* but no proper vector incrimination evidence has been obtained to support these notions. Kirk and Lewis (1955) suggested that *P. martini* as the vector of visceral leishmaniasis in Kapeota area in southern Sudan, where *P. orientalis* was absent.

Despite the progress made in mapping the geographical distribution of phlebotomine sandflies of the Sudan (Fig. 2) (Kirk and Lewis, 1951), little information is available about their distribution of sandflies in different ecological habitats of the country. Mainly, the distribution of *P. orientalis* was found to be associated with *Acacia-Balanities* forests grow on the black cracking clay in the

6

savannah regions in the endemic areas of kala-azar in eastern and southern Sudan (Elnaiem *et al.*, 1998a; Thomson *et al.*, 1999). However, recently, *P. orientalis* was recorded in village habitats in Khartoum State, a semi-desert area (Hassan *et al.*, 2007).

1.4.2. Leishmaniasis in Sudan:

Sudan is considered as one of the most important areas of leishmaniasis in the world. All forms of leishmaniasis i.e. CL, MCL, VL and PKDL occur in Sudan. These forms of the disease cause serious health problems and economic loss in the country such as disability of affected individuals and the cost of treatment, especially as most of those with leishmaniasis are on low incomes and live in rural areas (Fig. 3 distribution of leishmaniasis in Sudan).

Visceral leishmaniasis (VL) is considered to be one of the most important endemic diseases in the Sudan. The disease is mainly endemic in a wide belt extends from the eastern part of the country near the Ethiopian border, Blue Nile State, the White Nile in the Central Sudan and to the Upper Nile States with scattered foci in Kapeota in southern Sudan, Nuba Mountain and Darfur (Zijlstra & El-Hassan, 2001a).

The first case of VL in Sudan was described by Neave in 1904 in eight years boy from southern Sudan (Zijlstra & El-Hassan, 2001a). Occasionally, severe out breaks threatened the lives of many people in endemic areas occurred in Sudan. The first major epidemic was reported by Stephenson in 1940 in the Upper Nile Province, Southern Sudan where 300 cases occurred with death rate of 80% (Osman *et al.*, 2000; Zijlstra & El-Hassan, 2001a). Currently, during 1984-1994 epidemics, the disease has resulted in the death of 100,000 people out of population of 300,000 in relatively small area in the Western Upper Nile of Southern Sudan (Seaman *et al.*, 1996a). Similar reports indicate 1000 to 1300 cases of VL occur each year in Gedarif state, eastern Sudan (Elnaiem *et al.*, 2003).

Visceral leishmaniasis (Kala-azar, VL) is caused by *L. donovani senso lato* in eastern and southern Sudan (Oskam *et al.* 1998: Zijlstra & El-Hassan, 2001a) and transmitted by *P. orientalis* (Elnaiem *et al.*, 1998a; Lambert *et al.*, 2002; Hassan *et al.*, 2004, 2008). However, in the Kapeota area in South Sudan, where *P. orientalis* is absent, *P. martini* may be the main vector of VL (Miniter *et al.*, 1962). The Nile Rat (*Arvicanthis niloticus*) has been incriminated as the reservoir host for VL in Sudan (Hoogstraal & Heyneman, 1969; El-Hassan *et al.*, 1995). Also Hoogstraal & Heyneman, (1969) reported the rodent *Acomys aligena*, the spiny mouse and two species of carnivores: *Genetta g. senegalensis*, the Senegal Genet and *Felis serval phillipsi*, the Sudanese Serval Cat to be infected with leishmania parasite. However, recently the Egyptian mangoose, the Nile rat, the *Mastomys natalensis* and the

7

domestic dogs have been found infected with *L. donovani* in eastern Sudan (Elnaiem *et al.*, 2001; Deruere *et al.*, 2003; Hassan *et al.*, 2009).

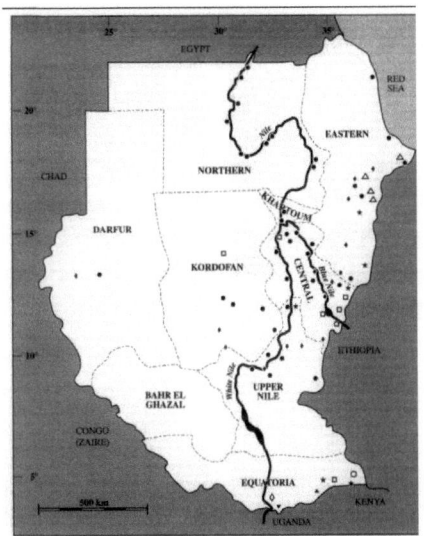

Figure 2: Distribution of *Phlebotomus* **sandflies in Sudan (El-Hassan & Zijlstra, 2001a).** O *P. alexandri*, Δ *P. bergerotti*, □ *P. duboscqi*, ◊ *P. longpipes* * *P. martini*, ◙ *P. oriental*, ● *P. papatasi*, ▼ *P. pedifr*, ▲ *P. saevus*.

Post-kala-azar dermal leishmaniasis (PKDL) is a complication of VL; it is characterized by a macular, maculopapular and nodular rash in a patient who has recovered from VL and who is otherwise well (Zijlstra *et al.*, 2003). PKDL was first described in Sudan by Christopherson in 1921 after he treated a VL patient (Zijlstra & El-Hassan, 2001b). Occasionally, PKDL may develop in the absence of VL history or during treatment of VL but more frequency it appears after 6 months treatment (Zijlstra *et al.*, 2003).

Cutaneous leishmaniasis is endemic in many regions of the country especially in the northern Sudan. The disease is characterized by present of skin papules, nodules, or noduloulcerative lesions, mainly on the exposed parts of the skin (El-Hassan and Zijlstra 2001a). The CL is caused by *L. major* (Abdallah *et al.*, 1973) although recently, *L. archibaldi* was isolated from CL lesions (Elamin *et al.*, 2008). The disease is transmitted by *P. papatasi* and the Nile rat *Arvicanthis niloticus* is believed to be the reservoir animal in the Khartoum area (El-Safi & Peters, 1991).

8

The first CL was first reported in Sudan by Thomson & Balfour in 1910 in two Egyptian soldiers who had contracted the disease in their country (El-Hassan & Zijlstra, 2001a). However, the first described CL case originating in the Sudan from the Nuba Mountains was described by Archibald in 1911 (Osman *et al.,* 2000; El-Hassan & Zijlstra, 2001a). Then the disease was reported from different parts of the country (Osman *et al.,* 2000; El-Hassan & Zijlstra, 2001a). In 1988 a major epidemic occurred in large parts of northern Sudan, causing thousands of cases (El-Safi *et al.,* 1991).

Sudanese mucosal leishmaniasis (SML) is a chronic infection of the upper respiratory tract and/or oral mucosa, caused mainly by *L. donovani* complex. The disease occurs in areas of the country endemic for VL, particularly among Masaleet and other closely related tribes in western Sudan (El-Hassan & Zijlstra, 2001b). The disease may develop during or after an attack of visceral leishmaniasis, but in most cases it is a primary mucosal disease (El-Hassan & Zijlstra, 2001b).

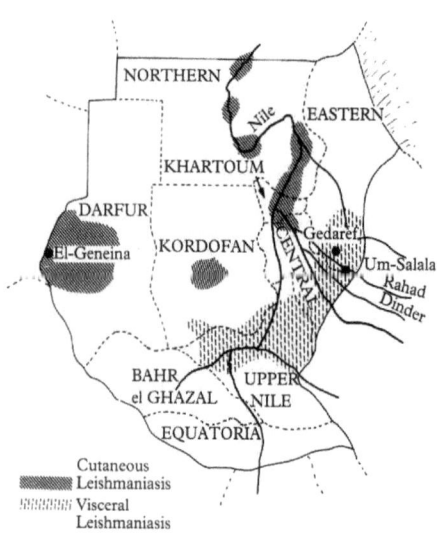

Figure 3: Distribution of leishmaniasis in Sudan (El-Hassan & Zijlstra, 2001)

1.5. Control of leishmaniasis and sandflies:

Control measures of leishmaniasis are the result of breaking one or more elements in the life cycle. There are many methods that can be used for control of leishmaniasis however, one method may be successful in one endemic area but not in another. Some of these measures target the reservoir such as occurred in former USSR, Jordan and Tunisia by eliminating the rodents by destruction of burrows and/or using rodenticides (Ashford, 1996; Klaus *et al.,* 1999).

The sand fly vector has continuously been the target for control measures. This included the destruction of breeding sites by removing garbage and debris left near houses and by covering cracks in buildings. In addition, spraying of residual insecticide inside houses and outside under windows were used (Llanos-Cuentas *et al.,* 2000). Plants like *Bougainvillea glabra* were shown to decrease the risk for leishmaniasis by reducing the life span of sand flies (Schlein *et al.,* 2001). Impregnated bed nets with various insecticides such as Deltamethrin were applied as control measure with significant reduction in leishmaniasis incidence rate (Alten *et al.,* 2003).

The human host was also a means for control either by allowing the patients to be treated 2-3 weeks after the appearance of the lesion, as is the policy in Jericho, to allow immunity to develop or by leishmaniazation as in Iran (Khamesipour *et al.,* 2005). Further, there are attempts to develop *Leishmania* vaccine but no definite results have been obtained yet (Valenzuela *et al.,* 2001).

1.6. Identification of sandfly vectors:

1.6.1. Morphological identification:

Identification of an insect vector and its differentiation from other insects is of a great importance in the field of medical entomology. Morphological identification is time consuming, sometimes problematic or impossible, particularly for closely related species which lack suitable differentiable morphological characters. However, many morphological identification keys were constructed for identification of New and Old World sandflies as well as sandflies of Sudan. Identification becomes more problematic when the insects are small in size, as in the case of sandflies. The only reliable characters were observed in males *Lu. longipalpis* bearing tergite one-spot and two spots from the coastal area of Mexico and Brazil (Mangabeira, 1969; Ward *et al.,* 1985, 1988). The variation among this population suggests presence of species complex.

1.6.2. Cross-matting experiments:

The occurrence of sibling species among sandflies was reported from several cross-mating studies which have been conducted within the Old and New World sandflies. Cross-mating between populations with different male sex pheromones was found unsuccessful (Ward *et al.,* 1988). Similarly, the crossing between Old World closely related sandfly species of subgenus *Phlebotomus* such as between *P. duboscqi* and *P. papatasi* (Ghosh *et al.,* 1999) and *P. bergeroti* and *P. papatasi* (Fryauff &

Hanafi, 1991) was unsuccessful. Moreover, cross-mating between *P. sergenti* specimens from Turkey and Israel was found successful. The crossing between one and two-spot populations of *Lu. longipalpis* in some cases produced an intermediate phenotype, a small spot on the third tergite (Ward *et al.,* 1988) as reported in some natural populations which indicates the existence of an intraspecific polymorphism in those areas (Ward *et al.,* 1988; Mutebi *et al.,* 1999). However, reproductive isolation in laboratory crosses and pheromone differences between one- and two-spot sandflies indicated absence of intermediate forms reflect the coexistence of two sibling species (Ward *et al.,* 1988).

1.6. 3. Cytogenetic analysis:

This method have been widely used in studies of taxonomy and population genetics of dipteran vector based on the detection of difference in a chromosome character such as banding patterns of the polytene chromosomes. These chromosomes can be found in either the fourth instar larvae or in the ovaries of semi/half gravid females. Only few cytogenetic studies were conducted on sandflies (Lanzaro & Warbrug, 1995). Studies on polytene chromosomes of *Lu. longipalpis* larvae salivary glands were described by White and Killick-Kendrick (1976). Moreover, the mitotic karyotype ($2n = 8$) which were conducted on larval cell brain of sandflies *Lu. longipalpis* from different geographic regions showed that there are differences in G-banding pattern and/or the position of the centromere on chromosome four which distinguished four putative sibling species (Yin *et al.,* 1999; Jiménez *et al.,* 2001). The disadvantages of this method include the requirements of expertise to carry out the assay, it is limited to early fourth instar larvae or semi-gravid females and it cannot be used to identify large numbers of samples (Gale & Crampton, 1987).

1.6.4. Enzymatic identification:

Isoenzyme analysis is considered as the 'gold standard' and reference method for identification and classification of species and strains, and for studying variability within cryptic vector species (Russell *et al.,* 1999). The method is a protein electrophoresis based on the migration of proteins under electrical field in a support medium (agarose gel or acetate cellulose membrane) (Hartl, 1988). Detectable proteins include functionally similar forms of enzymes designated as isozymes (Murphy *et al.,* 1990). Allozymes are different forms of the same enzyme resulting from allele variation (Crozier, 1993). In the isoenzyme analysis often between 10 and 20 isoenzymes are utilized.

Isoenzyme analysis of Diptera, such as mosquitoes, black flies and sandflies, has been used widely in the population genetic studies and the identification of different species (Maingon *et al.,* 2003). The useful of isoenzyme analysis for distinguishing between species of sandflies was first demonstrated by Miles and Ward (1978) by studying two different populations of *Lu. flavisculated* since then, the technique has been used successfully for taxonomic differentiation of Phlebotomine sandfly species (Ward *et al.,* 1981). Protein electrophoresis analysis of *Lu. longipalpis* has demonstrated substantial intra species variability at local population levels (Munstermann *et al.,* 1998) as well as substantial genetic differentiation among colonies originated from different geographic regions (Lanzaro *et al.,* 1993).

The advantage of the isoenzyme analysis technique over cytogenetic analysis is that it does not require a specific sex or larval stage (Miles, 1978). On the other hand, the method has disadvantages such as, it is expensive, slow and laborious and the sample must be fresh or frozen.

1.6.5. Molecular identification:

1.6.5.1. Polymerase Chain Reaction (PCR):

The polymerase chain reaction (PCR) is a widely used technique in molecular biology. It is used to amplify a piece of DNA by *in vitro* enzymatic replication. As PCR progresses using two sets of oligonucleotide primers, the DNA template is exponentially amplified. With PCR it is possible to amplify a single or few copies of a piece of DNA cross several orders of magnitude, generating millions or more copies of the DNA piece. The technique was first developed in 1983 by Kary Mullis (Bartlett & Stirling, 2003). PCR is now a common and often indispensable technique used in medical and biological research laboratories for a variety of applications (Saiki *et al.,* 1988).

Polymerase chain reaction (PCR) (Mullis & Faloona, 1987) has revolutionized molecular systematic and population genetic studies. This technique now uses a thermo stable bacterial polymerase extracted from the bacteria *Thermas aquatics* named as *Taq* polymerase to replicate DNA. PCR reduce the amount of the DNA used and a specific primers permit amplification of specific regions of the genome.

The PCR method to distinguish members of the sandfly species and complex are not common where only few studies were conducted (Muccio *et al.,* 2000; Testa *et al.,* 2002). This method was based on species-specific nucleotide sequences in the ribosomal DNA (rDNA) intergenic spacer regions which are useful in identifying species within the complex regardless of life stage and sex using either extracted DNA or fragments of a specimen.

1.6.5.2. PCR-based techniques:
1.6.5.2.1. Random Amplified Polymorphism DNA (RAPD):

Random-amplified polymorphic DNA (RAPD), also known as arbitrarily primed PCR that represents a powerful tool for identification of species, strains and the estimation of genetic variability in populations (Williams *et al.,* 1990). RAPD-PCR is based on its amplification of repetitive regions but amplifies many unique regions as well. The technique amplifies random fragments of the genomic DNA by using single 9-10 base-pair long primer of arbitrary nucleotide sequence with a minimum GC content of 60%. It is considered to be a fast, easy, unexpensive and yet very informative method (Diakou & Dovas 2001). During RAPD–PCR, the primer is annealed to the genomic template DNA at low temperature (35-39 °C) but also hold true at annealing temperature as low as 15 °C. These primers bind to homologous sequences along the genome and the PCR amplification only occurs where opposing primer site are 3000 bp apart. Hence, the single oligonucleotide can prime replication in both the forward and reverse direction.

RAPD genetic markers have also proved to be useful for a wide variety of molecular ecological applications. In the New World sandflies, RAPD was successfully used to discriminate between two closely related species, *Lu. youngi* and *Lu. spinacrassa* (Adamson *et al.,* 1993). Also, the RAPD distinguished successfully between different laboratory populations of *L. longipalpis* complex (Balbino *et al.,* 2006) and distinct geographic populations of *L. whitmanni* complex (Margonari *et al.,* 2004). However, only few studies using RAPD were performed on the Old World sandflies. RAPD was successful when has been used to compare genetic variation within and between five *Phlebotomus* species sympatric in southern Spain (Martin-Sanchez *et al.,* 2000). Also RAPD was used to develop species specific diagnostic profiles of *P. papatasi* and *P. duboscqi* (Mukhopadhyay *et al.,* 2000).

1.6.5.2.2. Single Strand Confirmation Polymorphism (SSCP):

Single Strand Confirmation Polymorphism (SSCP) is the simplest and most used method of mutation detection. PCR is used to amplify the region of interest and the resultant DNA is separated as single-stranded molecules by electrophoresis in non-denaturing polyaccrylamide gel (Orita *et al.,* 1989). A strand of single-stranded DNA folds differently from each another if it differs by a single base. It is believed that a mutation-induced change of tertiary structure of the DNA result in different motilities for the two strands. These mutations are detected as the appearance of new bands on auto radio grams (radioactive detection) by silver staining of bands or the use of fluorescent PCR primers which are subsequently detected by an automated DNA sequence.

The sensitivity of SSCP depends on the conditions e.g. temperature and ionic

13

environment that cause tertiary structure of single stranded DNA. Mutation detection for PCR-SSCP is generally high, (> 80%) in a single run for fragments shorter than 300 bp (Hayashi & Yandle, 1993). The sensitivity of PCR-SSCP decreases with the increasing in fragment length <300 bp being the optimum. For mutation detection in longer fragments (DNA >300bp) overlapping short primer sets can be used, or long PCR products digested with appropriate restriction enzyme prior to SSCP.

SSCPs are allelic variants of inherited genetic traits that can be used as genetic markers. However, SSCP analysis can detect DNA polymorphisms and mutations at multiple places in DNA fragments (Orita *et al.*, 1989). As a mutation scanning technique, SSCP is more often used to analyze the polymorphisms at single loci, as used in populations of ticks, leaf hoppers, mosquito and closely related endo-parasitic wasps (Hiss *et al.*, 1994).SSCP was used to detect polymorphism of SP-15 in *P. papatasi* sandflies(Elnaiem *et al.* 2005).

The only major disadvantage associated with the SSCP analysis is the requirement for sequence information with which to construct specific primer to amplify unique mitochondria and nuclear genes.

1.6.5.2.3. Restriction Fragment Length Polymorphism (RFLP):

Restriction Fragment Length Polymorphism (RFLP) is also a characterization technique that is based on fragment size chosen by restriction enzyme (Collins *et al.*, 1987). PCR product or DNA is digested with one or more restriction enzymes and the resulting fragments are separated according to molecular size using gel electrophoresis. RFLP can be used to separate several cryptic insect species. The internal transcribed spacer region 1 (ITS1) of the nuclear ribosomal DNA genome is a useful marker for diagnostic purpose. This diagnostic procedure can readily identifying alcohol preserved individual adults, juveniles and eggs of a species in 24-48 hour. Currently, sandflies have been successfully identified to species level by restriction-fragment-length polymorphism (RFLP) analysis using the products of the PCR-based amplification of the flies' 18S ribosomal RNA (rRNA) genes (Aransay *et al.*, 1999; Barroso *et al.*, 2007). The technique is sensitive enough that a PCR-RFLP can be generated from as a little as a single insect leg. RFLP analysis is highly accurate and can be used regardless of the live stage or sex of the insect tested but the major disadvantage is the relatively high cost of the endonuclease enzymes (Copeland *et al.*, 1992).

1.7. Genetic variation among sandfly vector:

Despite of the medical importance of sandflies, the study of their genetics has remained neglected (Lanzaro & Warburg, 1995). However, levels of genetic diversity using genetic markers were detected among *L. longipalpis* the vector of leishmaniasis

14

in Latin America species (Arrivillaga *et al.*, 2002; Maingon *et al.*, 2003). The suggestion that *L. longipalpis* is a cryptic species complex raised serious concern of species identification and their roles in the transmission of leishmaniasis.

The morphological variations in *Lu. longipalpis* in the eastern Brazil were the first evidence cited for the presence of cryptic species with in a species complex (Lane, 1968). Then several phylogenic analyses of biochemical and molecular characters revealed the occurrence of complexes among some species *Lu. longipalpis* (Lanzaro *et al.*, 1993), *Lu. whitmani* (Ishikawa *et al.*, 1999) and *Lu. shannoni* (Cárdenas *et al.*, 2001). For such studies a gene of high mutation rates were chosen. Although the cytochrome b gene has been shown to be useful at the subgeneric taxonomic level for *Larroussius* (Esseghir *et al.*, 1997), the internal transcribed spacer 2 (ITS2) of the ribosomal DNA has provided resolution in three pilot studies. Two of these studies were done for the subgenus *Larroussius* (Mancini *et al.*, 1997) and the third was on the subgenus *Paraphlebotomus* (Depaquuit *et al.*, 2002). However, many studies incriminated siblings among sandfly populations based on morphological differences (Mangaberira *et al.*, 1969), behavioral differences (Lainson, 1988) or mating non compatibility (Ward *et al.*, 1988).

The banding patterns of the polytene chromosomes are constant corresponding to the gene sequence. Thus any change in gene sequence might be detected by direct comparison of chromosomes from different individuals. Taxonomists have been able to look at the genotype of the species rather than the phenotype after the discovery of polytene chromosomes in many of the Diptera, specifically in the sandflies (White and Killick-Kendrick, 1976). The examination of banding patterns of these chromosomes has been used extensively for verifying the existence of cryptic species (Yin *et al.*, 1999; Jiménez *et al.*, 2001) and /or to look at intra specific variations (Pape, 1992).

DNA sequences and allozyme loci also provide two independent genetic markers by which the cohesion of morphologic taxonomic grouping may be tested. An examination of the genetic relationships was done among 20 species of *Lutzomyia* and *Brumptomyia* using nine allozyme loci and the last 285 base-pairs of the mitochondrial cytochrome b gene. A previous study suggested the topologies of trees based on this gene reflect the phylogenetic relationship between species rather than populations (Torgerson *et al.*, 2003).

However, in the transmission of different forms of leishmaniasis, phlebotomine species form transmission cycles with specific associations between *Leishmania* parasites and vertebrate hosts and unique epidemiologic patterns. Thus, the genetic relationships between *Lutzomyia* vectors and non-vectors species seem fundamental for understanding the evolution of zoonotic transmission cycles.

1.8. Phylogenetic Trees:

Phylogenetic trees are powerful means for summarizing evolutionary relationships (Yang, 1995). Many different criteria can be used to infer phylogenetic trees from morphologic or molecular data. Phylogenetic trees can be used to detect relationships among copies of a gene or among loci of a multigene family (Croan, 1997). Several methods for estimating phylogenetic trees are available. Some of the more commonly used methods include neighbor joining (NJ) (Saitou & Nei, 1987), maximum likelihood (ML) and maximum parsimony (MP) (Felsenstein, 1981).

NJ analysis yield a point estimate of a minimum evolution tree based on data transformed into a pairwise distance matrix. In this method, the alogrithim for finding the tree and the criteria for assessing its quality is combined.

ML analysis uses an explicit model of evolution (direction and character change) to derive the tree most likely to have occurred given in the data to that many different trees can be built and tested.

MP is a criterion for selecting an optimal tree based on the principle that the tree requires a minimal number of changes among methods for testing the internal statistical support for the tree is bootstrap.

The bootstrap is re-sampling rebuilds a number (usually 1000) of model data sets based on the test set (by sampling with replacement) and re analyzing them with the chosen criteria. The retention of nodes in the set of bootstrap trees is a strong indication of their robustness. Bootstrap values 65% are considered as robust.

1.9. Objectives:
1.9.1. General objectives:

The present study was carried out using molecular tools to identify *P. orientalis* and to investigate the inter- and intra-species genetic heterogeneity in the populations of *P. orientalis* from different geographic regions of Sudan.

1.9.2. Specific objectives:

In specific the present study was designed;
1. To evaluate molecular-based technique (mtDNA-PCR) for identification of *P. orientalis*.
2. To determine the genetic heterogeneity in *P. orientalis* from different geographic locations (northern and eastern Sudan) using RAPD-PCR.
3. To analyze the genetic structure of different populations of *P. orientalis* using appropriate software.

Rationale:

Behavioral and morphological differences in *P. orientalis* varied geographically (field observation; Hassam M.M. per comm.), suggested the existence of species complex among this species in eastern Sudan. However, no genetic taxonomy for the phlebtomine was done to incriminate species complexes among the vectors of leishmaniasis in Sudan. Therefore, molecular investigation is needed to verify genetic differences among and within the population of the *P. orientalis* in different geographic regions in eastern and northern Sudan. The study of the genetic structure might elucidate the extent of variation estimate degrees of relatedness among population and determine the number of sibling within the study area which will help to understand the epidemiology of the transmission of different strains of *leishmania donovani* in eastern Sudan. Besides, it may verify the transmission capacity of some *phlebtomus* sandflies in Sudan. Such data will add valuable information to the epidemiology of transmission of leishmaniasis in the country which in turn will lead to set suitable control strategies.

CHAPTER TWO

MATERIALS AND METHODS

2.1. Collection sites of sand flies:

The field work of this study was carried out to collect *P. orientalis* from three areas in eastern Sudan (Gedaref State); Dinder National Park(DNP), Rahad River area (RH) and Atbara River area (ATB) and two areas in northern Sudan; Surogia village (Khartoum state) and White Nile area(White Nile State). Three field trips were carried out in each site during the period 2006-2007. The sandflies in each village were collected during the dry season (January – early July).

2.1.1. Eastern Sudan:

The selected sites for collection of sandflies in eastern Sudan were Dinder National Park (DNP; 35° 2' E, 12° 36' N), Rahad River area (RH; 35° 11' E, 12° 51' N) and Atbara River area (ATB; 36° 13' E, 14° 15'). The ecology of the area in eastern Sudan has been described by Elnaiem *et al*. (2001, 2003). In brief, the land is a flat, but in many places it is interrupted by seasonal rivers, khors and few ground surface water collecting depressions. The soil is mainly chromic vertisol soil (black cotton soil), interrupted with few sandy and silts soils known as azaza.

The climate prevailing in the area is tropical continental with annual rainfall 800-1000 mm. The year is divided into a hot dry season (summer: March-May), a warm rainy season (autumn: June-October), and a moderately warm winter (November- February). The mean temperature and relative humidity of the area are 32.3 °C and 25.9 % in the summer, 26.2 °C and 54.4 % in autumn and 27.°C and 29.8% in winter.

Vegetation in the area consists largely of mixed species that dominated by *Acacia seal* (Taleh), *Balanites aegyptiaca* (Hijlij) and *Zyzyphus spina Christi* (Sidir). The villages are partially cleared of the natural vegetarian, dominated by *B. aegyptiaca* and *Acacia* trees within the villages and *Zyzyphus spina Christi* on the river bank. The villages in surrounded by fields of sorghum, sesame, Dokhun (*Pennisetum typhodium*) and Groundnuts (*Arachis hpogaea*). The vegetation in the DNP is similar to that in the villages, but the vegetation in the park which is protected by environmental conservation laws, is densely covered with mature forests of *Acacia, Balanites, Combretum* and many other tree species. The floor of this woodland is occupied by *Sorghum spp., Pennisetum rumosum, Setaria sp,* and *Aristida plumosa*.

The DNP contains variety of large wild mammals including Reedbuck

18

(*Redunca redunca*), Oribi (*Ourebia ourebi*), Water buck (*Kobus defassa*), Roan antelope (*Hippotragus equinus*), Bush buck (*Tragelaphus scriptus*), Warthog (*Phacochoerus aethiopicaus*), buffalo (*Syncera scaffer*), Greater kudu (*Tragelaphus angasi*), Lion (*Panthera leo*), Hyaena (*Crocuta crocuta*), Grivet monkey (*Cercopithecus aethiops*), Patas monkey (*Erythrocebus patas*), Baboon (*Papio anubis*). There are also many colourfull birds, such as starlings (*Lamprotornis spp*), bee eater (*Merops spp*), herons (*Ardea melanocephala*), egrets (*Egretta alba*), marabou stork (*Ciconia episcopus*) and the Abyssinian Roller (*Coraccias abyssinica*) (Dasmann, 1972; Ernst and Elwsila, 1985). However, in the villages, most of the wild animals were eliminated by pouching and replaced with domestic animals like sheep, cattle, goats and dogs although many rodents and few wild cats can be seen.

The villages have diverse human populations that are dominated by West African ethnic groups who have recent history of settlement in the area. They came at the early 1960s as labourers in the mechanized agricultural scheme. The majority of people came to the area, during the drought of 1984, which struck western Sudan (Elnaiem *et al.*, 2003). The total population in the area are approximately about 200 000 people. The people of the area belong to Hausa tribes who migrated from northern Nigeria (West Africa), Masaleet, Fur, Gemer, Meema and Miseryia from western Sudan and Chad in West Africa whereas, Bin Amir and Arab people from eastern Sudan. Also, there are few people belong to Nuba tribes from Nuba Mountain and Dinka people from southern Sudan. The villagers live in typical African huts constructed mainly of wood and bamboo and thatched with grass. A few huts in certain village are constructed of mud layers and are roofed with wood and bamboo thatched with grass.

The Dinder National Park (DNP) is protected by environmental laws. Therefore, the park is completely uninhabited by villagers except for five, widely dispersed camp for the Game Wardens which inhabited by 150 wardens (20-40 in each camp) who live in huts similar to those in tow villages.

2.1.2. Northern Sudan:

Two sites for sandflies collected were selected in northern Sudan these were; Surogia village (Khartoum State) and White Nile area (White Nile State).

Surogia village is located on the eastern bank of the River Nile (32° 14′ E, 14° 72′ N) 30 km form Khartoum. It is lies in the endemic zone of cutaneous leishmaniasis. The ecology of the area was described by El-Sayed *et al.* (1991). Generally the area is flat and covered by alluvium of still clay and sand deposited by the river. Being near the Sahara desert, Surogia area experiences climate with three distinct seasons, winter (November-February), summer (March-June), and autumn (July-October). Vegetation in the area is of the desert scrub type dominated by

Acacia trees. Majority of the animals in the village are domestic animals like sheep, cattle, goats and dogs. Moreover, few wild animals such rodents and few wild cats can be seen nearby. The history of the residence in the village is back to Mahadyia period. The people belong to Arab tribes with few emigrated people from southern Sudan. The people of the village live in houses with large rooms (4X5 m^2) constructed of mud layers or bricks, roofed with grass and mud or corrugated iron sheets. People spend the night indoors during the winter and most of autumn. In the summer and hot days of autumn, they usually sleep outdoors overnight.

White Nile area site is located on western bank of the White Nile (32 $^\circ$ 14' E, 14° 72' N), 200 km southern of Khartoum. The area is flat and covered by an alluvium of silt clay soil of the two types, black cracking clay soil and sand deposited by the river. The area resemble a semi desert region therefore, it is part of the arid climate with three distinct seasons; winter (November-February), summer (March-June) and autumn (July-October) with an estimated annual rainfall of <50 mm. the area is dominated by the desert vegetation which is mainly *Acacia* spp. trees Majority of the animals in the village were domestic animals like sheep, cattle, goats and dogs. The people belong to Hassaniya Arab tribes.

2.2. Phlebotomine sandflies collection, preservation and identification:

Sandflies were collected from the above mentioned sites using sticky paper traps and CDC miniature light traps as described by Quate (1964), Lewis (1973) and Killick-Kendrick (1987).

2.2.1. Sticky paper traps:

These traps were Xerox paper sheet (15 × 21cm^2) coated on both sides with diesel oil and fixed vertically on sticks held at 15 cm above the ground level. Then oil traps were fixed between 18:00-06:00 HR and collected by the next morning. Sandflies captured by these traps on the oiled papers were removed using small brushes, washed in dilute detergent and preserved individually in 70% alcohol in eppendorrf tubes for identification.

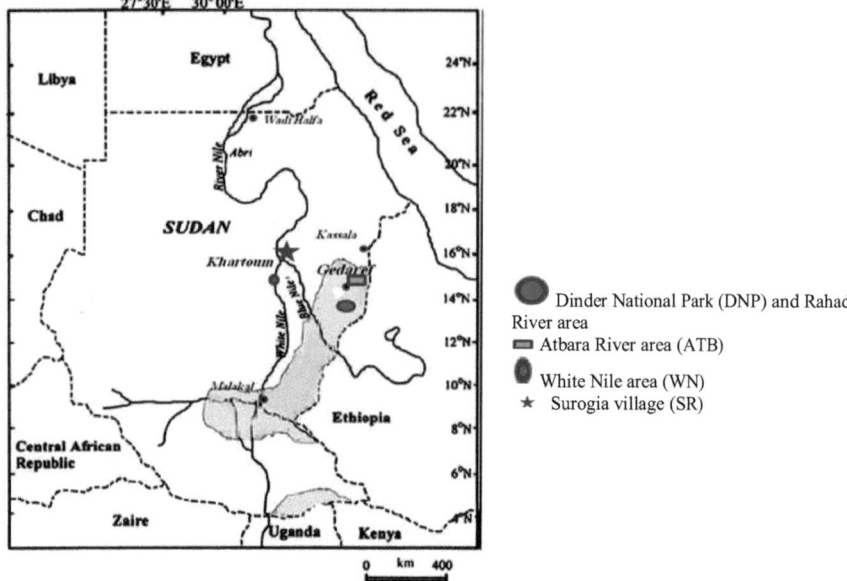

27°30'E   30°00'E

Egypt
Libya   24°N
                    Wadi Halfa   22°N
Red Sea
                    Abri   20°N
Chad   SUDAN   Kassala   18°N
            Khartoum   Gedaref   16°N
                                          14°N   ● Dinder National Park (DNP) and Rahad
                                          12°N   River area
                                                  ▭ Atbara River area (ATB)
            Malakal   10°N
                        Ethiopia   8°N   ● White Nile area (WN)
Central African   ★ Surogia village (SR)
Republic   6°N
                                           4°N
Zaire   Uganda   Kenya

0   km   400

Figure 4: A map showing *Phlebotomus orientalis* **collection sites from different geographic regions in Sudan**

2.2.2. Light traps:

To collect nocturnal alive sandflies, CDC miniature light traps (Model 512, John Hock Co., Gainesville, Florida, USA), connected to 6volts rechargeable batteries and sandflies cages were used. Light traps were fixed at 20 cm above the ground level at outdoor and indoor sites between 18:00-06:00 HR. Captured sandflies were separated from other insects and preserved individually in 70% alcohol in eppendorrf tubes for identification.

2.3. Microscopic identification of *Phlebotomus orientalis*:

In the laboratories before applying any analysis, the head and the two last segments of females sandfly (specimens preserved 70% ethanol) were carefully mounted sin Berlese's medium (homemade medium) on glass slide, as described by Killick-Kendrick and Killick-Kendrick (1987). Using a pair of fine dissecting needles, under a dissecting microscope, the two last segments and the head were separated from the rest of the body, placed on the same slide and were covered with a cover slip. The slides were then left for 12 hr to air dry. The prepared specimens were identified using binocular microscope at 40X. The female specimens were identified using a proper identification keys constructed by Kirk and Lewis (1951),

21

Quate (1964) and Abonnenc and Minter (1965). The main features used for female sandflies identification were the spermatheca, the pharyngeal and the cibarial toothed structures.

2.4. Molecular techniques for Identification of *Phlebotomus orientalis*:
2.4.1. DNA extraction:
2.4.1.1. Phenol/chloroform method:

DNA from *P. orientalis* specimens was extracted using phenol/ chloroform method as described by Mortz and Hillis (1987). Briefly, the ethanol-preserved female sandflies were individually grinded into fine powder using glass pestle in eppendorrf tube. The tissues were placed in 500µl of STE buffer (0.1MNaCl, 0.5MTris-HCl Ph7.5, 0.001EDTA) in a 1.5 tube. Then 25 µl of 10mg\ml stock of proteinase K were added in STE and mixed well. After that 25µl of 20% SDS were added and incubated at 37 °C overnight. Next morning, an equal volume of PC (solution of Phenol, Chloroform and Isoamyl alcohol in ratio of (25:24:1). was added, mixed gently and then incubated at room temperature for 5 minutes. Then the samples were centrifuged for 5 minutes at 7000 rpm. The aqueous layers were carefully removed with micropipette and transferred to clean eppendorrf tubes. After that an equal volume of CI (Solution of chloroform and isoamyl alcohol in ratio of 24:1) was added, mixed gently and then incubated at room temperature for 3 min. The samples then were centrifuged for 5 min at 7000 rpm. The aqueous layers were carefully removed with micropipette and transferred to clean eppendorrf tubes. At the end 45µl of 2M NaCl and 1ml of cold absolute ethanol were added to the DNA precipitates. Then the mixtures were incubated overnight at -20 °C. The DNA precipitates were then spinet down in a micro centrifuge tubes for 20 min at 7000 rpm. Ethanol was decanted and the DNA pellets were dried. Finally the pellets were re-suspended in 20ml of PCR water and stored at 4 °C for subsequent PCR analysis.

2.4.1.2. Sodium Tris EDTA (STE) method:

DNA from *P. orientalis* specimens was extracted as described by Balbino *et al.* (2006). Briefly, the ethanol preserved females were placed individually into 1.5µl tube with 50µl grinding buffer (0.1mMNaCl, 0.1MTrisHCl pH8.0) in eppendorrf tubes. The samples were grinded to a fine powder using glass pestle, and then the homogenates were incubated at 95 °C for 30 min. The supernatants were transferred to new tubes and then 200µl of PCR water was added. The DNA solutions were stored at 4 °C until for subsequent PCR analysis.

2.4.2. Polymerase Chain Reaction:
2.4.2.1. Species-specific PCR:
2.4.2.1.1. Design of specific primers:

Members of subgenus *Larroussius* in the Sudan and the East Africa neighboring regions were short listed from the literature. Then the cytochrome b (Cytb) mtDNA sequences of *P. orientalis* were downloaded from the NCBI database as well as Genbank (Accession No. AF161203) (Esseghir *et al.*, 2000). In order to avoid any non-specific amplification of the non-targeted genera, the sequence obtained for *P. orientalis* from the database were subjected to CLUSTALX programme (Jeanmougin *et al.*, 1998). The regions showing variations among subgenera were picked up. These sequences were then subjected again to alignment where the maximum regions showing conservations were selected as genus specific primers using Primer3 software (http://fordo.wi.mit.edu/cgi.-bin/primer3/primer3_www. cgi; Rozen & Skaletsky, 2000). Genbank program BLAST (Altschul *et al.*, 1990) was used to further ensure that the proposed primers were complementary with the target species and not with the non-targeted one.

The sequences of the specific-primer designed for identification of *P. orientalis* Cyt b mtDNA, size of the PCR product and the annealing temperature used in this study were shown in table 1.

2.4.2.1.2. PCR assay:

The procedure described by Paskewitz and Collins (1990) was used in the amplification of *P. orientalis*. A total of 25 µl volume using 3 µl DNA solution and 300 pmol of each of the two primers (MM1R and MM1F) (Inqaba Biotec, South Africa) were used. The final concentration of the PCR mix contained 10 mM Tris HCl, pH 8.3, 1.5 mM $MgCl_2$, 50 mM KCl, 0.01% Triton X100, 200 µm dNTP each (vivantis, UK), and 0.5 µl (2.5 units) of *Taq* polymerases (vivantis, UK). The PCR (PCR machine, Biometra, Germany) conditions was 35 cycles, denaturation at 94 °C for 5 min, followed by annealing for 1 min at 72 °C, extension for 1 min at 72°C and extension for 10 min at 72 °C to complete DNA amplification. PCR products then were tested by electrophoresis in 1.5% agarose gel containing ethidium bromide. The PCR amplified products of 675 pb were identified as *P. orientalis*.

2.4.2.1.3. Agarose gel electrophoresis:

Specific PCR products were visualized using agarose gel electrophoresis as described by Sambrook *et al.* (1989). Briefly, 1.5g agarose was melted in 1X TBE (Tris, Boric acid, EDTA) and 2 µl of Ethidium bromide was added. Then the agarose gel mixure was poured in a horizontal apparatus covered with 1X TBE buffer. Then 5µl of each PCR product was mixed with 2µl of loading dye (bromophenol blue and water). A DNA ladder (100 pb molecular weight; vivantis, UK) was also loaded for the determination of the amplified DNA size. The apparatus was connected to the power supply of a voltage 80V for two hours. After completion of the run, the gel was observed under ultraviolet light to determine if PCR products have been successfully amplified. Photography of the gels was performed by gel documentation system (UVP-91786, USA).

Table 1: Common characters of the Species-specific primers used for identification of *Phlebotomus orientalis* from different geographic regions in Sudan

	Primer 1 (MM2F)	Primer 2 (MM2R)
Sequence	5'-tttactctctgctattccttatctagg -3'	5'-tctcagatttttgaaattagaggattt -3'
Molecular weight (Max/Min)	8157/8157	8317.9/8317.9
Conc. nmole	55.07	47.83
Optical Density (OD)	14.44	14.44
Melt TM (Max/Min)	61.57/61.57	57.01/57.01
GC% (Max/Min)	37.04/37.04	25.39/25.39

2.4.2.2. Random Amplified Polymorphic DNA (RAPD):
2.4.2.2.1. RAPD-PCR assay:

Thirty RAPD primers (Append 1), (Inqaba Biotec, South Africa) were used as genetic markers in this study to screen DNA from *P. orientalis* from five different geographic regions in Sudan. The primers were 10mer in length with G+C content of 60-70%. The amplification conditions were done as described by Williams *et al.* (1990). Amplification reaction was carried out in a final volume of 25µl containing 3µl DNA; 0.5µl dNTPs (200µM of each dNTP final concentration); 2.5µl 10X PCR buffer (50mM KCL; 1.2 Triton-x; 1.5 mM $MgCl_2$) (1x final oncentration); 2.0µl each primer (15 pmole of each primer as a final concentration); 0.2µl *Taq* polymerase (1 U of *Taq*). The thermocycler was programmed for 35 cycles with the following conditions: denature for 2 min at 94 °C, annealing for 1 min at 35 °C and extension for 2 min at 72 °C. The final extension was held for 10 min at 72 °C after the cycling was completed. The complete reactions were placed at 4 °C until gel analysis was accomplished. Table 2 shows the RAPD primers that produced amplification.

2.4.2.2.2. Gel analysis of RAPD:

Five microliters of the RAPD-PCR reaction products were mixed with two-µl of loading dye and loaded on to 0.8% agarose gel with electrophoresis buffer (TBE of 89 mMTrise-base, 89 mM boric acid, 2mM EDTA pH 8.3). 2µl of 10mg\ml ethidium bromide stain were directly added to the agarose before pouring the gel. DNA marker 100-3000 bp was used as a molecular weight ladder to determine the RAPD band sizes.

Gels electrophoresis was performed at 50v in 1X TBE buffer for three hours. The gels were put on the ultraviolet transilluminator and photographed using gel documentation system.

Table 2: **Sequence of the four RAPD primers that produced clear and reproducible DNA fragments for** *Phlebotomus orientalis* **from different geographic regions in Sudan.**

RAPD primer	The sequence
RAPD 1(r469)	TCGCAACGTC
RAPD 2(ι751.2)	GGGCACTCCG
RAPD 3(r807.1)	GCCTTCATCT
RAPD 4 (r564.1)	GCCTCCTACT

2.4.2.2.3. Reproducibility of RAPD primers:

The reproducibility of the RAPD primers used was assessed by documenting the repeatability of RAPD- PCR 2-3 times using DNA of *P. orientalis* from different areas.

2.4.3. Data analysis:

Data obtained from the RAPD banding pattern were analyzed by comparing the number of bands by each RAPD primer within and between the populations of the five geographic regions. Bands were scored manually as 1 (present) or 0 (absent) and only bright and reproducible bands were scored. Allelic frequency and genotype were calculated manually at each locus in each population. Dice genetic distance was calculated using XLSTAT software 2009, the geographic distances between the sample sites were calculated, so isolation by distance could be examined by Mantel test (1000 permutation) using Arlequine (ver3.11) software. Arlequin software was also used to calculate the genetic structure and the fixation index by AMOVA test. Phylogenetic trees were constructed using a neighbor-joining algorithm by RAPDistance 2.0 software and XLSTAT 2009 software.

CHAPTER THREE
RESULTS

3.1. Sand fly species of the study area:

The results microscopic identification of the sandflies collected from DNP, Rahad RH, WN, SR and ATB during November 2006 and November 2007 showed that eleven species of sandflies were recorded, five *Phlebotomus* species and six *Sergentomyia* species. These species were *Phlebotomus orientalis, P. papatasi, P. bergroti, P. roubaudi, P. rodhaini, Sergentomyia schwetzi, S. clydei, S. squamipleuris, S. antennatus, S. africana,* and *S. bedfordi*. Most of these species were detected in all the study sites investigated; the exceptions were *P. rodhaini* which was absent from WN, *P. roubaudi* which was absent from DNP & WN and *S. bedfordi* which absent from Atbara and WN. Moreover, *P. bergeroti* was recorded only in RH.

A total of 2535 female sandflies were captured using sticky paper and light traps. Of the total collections, *P. orientalis* was 18.03% (457 specimens), whereas *P. papatasi* (693; 27.34%), *S. clydei* (568; 22.40%) and *S. schwetzi* (491; 19.37%) were the most abundant species recorded in this study (see table 3). *Phlebotomus rodhaini* (12; 0.47%), *P. bergroti* (5; 0.20%) and *S. bedfordi* (8; 0.32%) were collected in very few numbers than other species (table 3).

3.2. Identification of *Phlebotomus orientalis* using PCR:
3.2.1. DNA extraction:

The STE and phenol/chloroform methods yielded high quality of *P. orientalis* DNA. The STE method was found to be less time consuming, cheaper and yielded more DNA compared to the phenol/chloroform method. When analyzed on agarose gels, the DNA extracted by the two methods was of good molecular quality with no smear.

3.2.2. Sensitivity of the designed primers:

The designed primers (MM2F and MM2R) from the cytochrome b (Cytb) mtDNA sequences of *P. orientalis* were found to be of high sensitivity (100%) to amplify the DNA of *P. orientalis*. However, these primers did not amplify DNA of other sandfly species such as *P. papatasi, S. schwetzi* and *S. clydei*.

Table 3: Numbers and percentages of sandfly species collected from different geographic regions in Sudan using light and oil traps during November 2006-Novemver 2007.

Sandfly species	Number (%) of sandflies collected from different sites					
	DNP	RH	WN	SR	ATB	Total
Phlebotomus orientalis	203(67.2%)	112(40%)	50 (25%)	55(9%)	37(3.2%)	457(18.03%)
P. papatasi	30(10%)	80(28%)	65(32.5%)	417(69%)	101(8.7%)	693(27.34%)
P. roubaudi	0(0.0%)	4(1.4%)	0(0.0%)	32(5.3%)	2(0.17%)	38(1.50%)
P. rodhaini	6(2.0%)	3(1.1%)	0(0.0%)	2(0.33%)	1(0.07%)	12(0.47%)
P. bergroti	0(0.0%)	5(1.8%)	0(0.0%)	0 (0.0%)	0(0.0%)	5(0.20%)
Sergentomyia clydei	26(8.6%)	18(6.5%)	30(15%)	34(5.6%)	460(39.7%)	568(22.40%)
S. antennatus	5(1.7%)	12(4.3%)	5(2.5%)	30(5.0%)	23(2%)	75(2.96%)
S. Africana	7(1.8%)	13(4.7%)	5(2.5%)	5(0.82%)	18(1.6%)	48(1.89%)
S. squamipleuris	10(3.3%)	16(5.8%)	25(12.5%)	3(0.5%)	86(7.4%)	140(5.52%)
S. schwetzi	15(5.0%)	14(5.1%)	20(10%)	10(1.6%)	432(37.2%)	491(19.37%)
S. bedfordi	4(1.4%)	2(0.8%)	0(0.0%)	2(0.33%)	0 (0.0%)	8(0.32%)
Total	306(12.07%)	279(11.01%)	200(7.89%)	590(23.27%)	1160(45.76 %)	2535(100%)

3.2.3. Species specific-PCR results:

Figure 5 shows the amplified DNA fragments of *P. orientalis* from different geographic regions in Sudan using species specific PCR. All DNA obtained from *P. orientalis* (200 specimens) showed DNA fragments at 675 bp by the species-specific PCR. However, all other sandfly species used as PCR controls in this study (*P. papatasi, S. schwetzi* and *S. clydei*), were not amplified with the same primers.

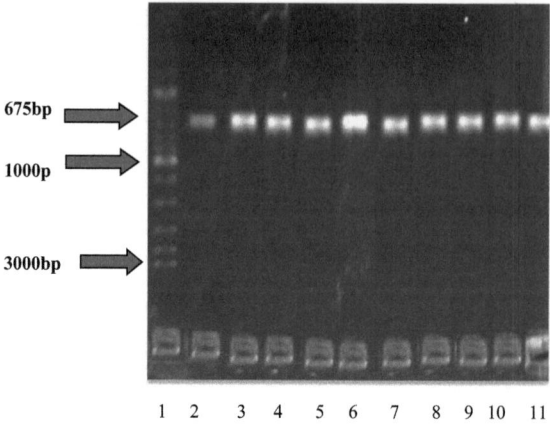

Figure 5: Species-specific PCR amplification of mtDNA of ***P. orientalis*** collected from the five geographic regions during November 2006-November 2007.

Lane 1: Molecular weight marker
Lanes 2-3: *P. orientalis* samples from DNP
Lanes 4-5: *P. orientalis* samples from RH
Lanes 6-7: *P. orientalis* samples from WN
Lanes 8-9: *P. orientalis* samples from SR
Lanes 10-11: *P. orientalis* samples from ATB

3.3. RAPD profile:

The RAPD-PCR amplification profiles of the 30 primers tested showed that only four RAPD primers (r469.1, r564.1, r751.2 and r807.1) produced a satisfactory level of variability and reproducibility (see fig. 6, 7, 8 and 9). The sizes of the produced bands (DNA fragments) and their numbers in each population are shown in table 4. In case of these four RAPD primers, 7–10 bands with sizes ranging from 100 to 3000 bp were obtained. Two DNA fragments one at 900 bp and one at 600bp were produced by r564.1 and r469.1 in one individual from DNP and one SR respectively. Also, DNA fragments at 100 bp and at 500 bp in three individuals from SR population were produced by r564.1 and r807.1 respectively. A 500 bp and 400 bp DNA fragments produced by r469.1 were not observed in ATB and WN populations respectively. The DNA fragment at 500 bp produced by r 564.1 was not observed in WN populations and the band size of 500 bp produced by r807.1 was not observed in ATB population. Moreover, two common bands at 300 bp were produced by r469.1 and r807.1 in all *P. orientalis* populations.

3.4. Estimation of genetic diversity within and among *Phlebotomus orientalis* populations from different geographical regions in Sudan:

3.4.1. Genotypes and alleles number produced by RAPD-PCR:

Table 5 shows the number of genotypes and the DNA fragments detected by the four RAPD primers in the populations of *P. orientalis* from different geographic regions in Sudan. The number of DNA fragments and genotypes detected were found to be varied according to the geographic origin of the sandflies collection. The DNP population produced the greatest number of genotypes (62) whereas, WN population produced lowest number of genotypes (9), although this population showed highest number of DNA fragments (717). In contrast, the ATB population had the lowest number of DNA fragments (112).

1 2 3 4 5 6 7 8 9 10 11 12 13 14

Figure 6: RAPD-PCR amplification obtained by r469.1 for *Phlebotomus orientalis* **collected from different geographic regions in Sudan during November 2006- November 2007.**

Lane 1: Molecular weight marker

Lanes 2-14: *P. orientalis* samples (A: DNP; B: RH; C: WN; D: SR and E: ATB)

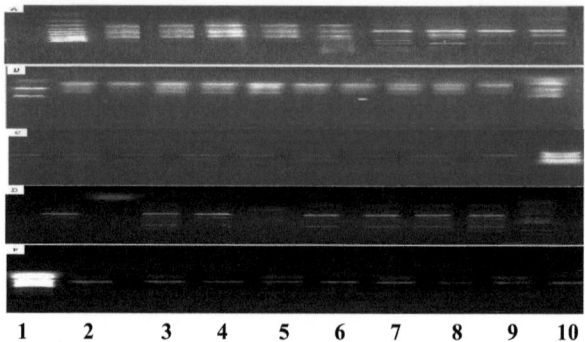

1   2   3   4   5   6   7   8   9   10

Figure 7: RAPD-PCR amplification obtained by r564.1 for *Phlebotomus*
orientalis **collected from different geographic regions in Sudan during**
November 2006- November 2007.

Lane 1: Molecular weight marker

Lanes 2-10: *P. orientalis* samples (A: DNP; B: RH; C: WN; D: SR and E: ATB).

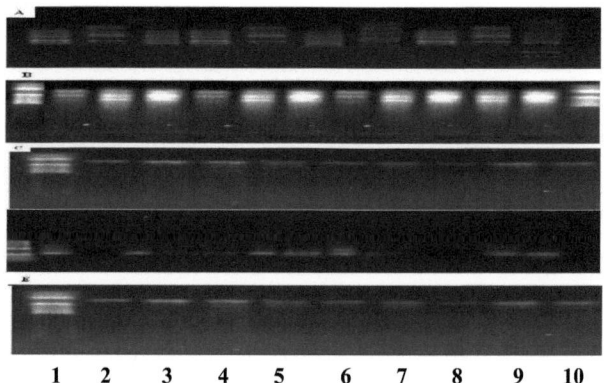

1   2   3   4   5   6   7   8   9   10

Figure 8: RAPD-PCR amplification obtained by r751.2 for *Phlebotomus*
orientalis **collected from different geographic regions in Sudan during**
November 2006- November 2007.

Lane 1: Molecular weight marker

Lanes 2-10: *P. orientalis* samples (A: DNP; B: RH; C: WN; D: SR and E: ATB).

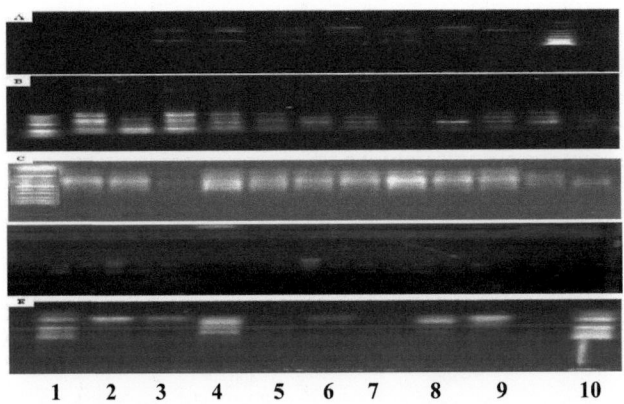

          1   2   3   4   5   6   7   8   9   10

Figure 9: RAPD-PCR amplification obtained by r807.1 for *Phlebotomus*
orientalis **collected from different geographic regions in Sudan during**
November 2006- November 2007.

Lane 1: Molecular weight marker

Lanes 2-10: *P. orientalis* samples (A: DNP; B: RH; C: WN; D: SR and E: ATB).

Table 4: DNA fragments sizes and their numbers produced by the four RAPD primers obtained for *Phlebotomus orientalis* from different geographic regions in Sudan.

Population of *P. orientalis*	r469.1		r564.1		r751.2		R807.1	
	# frag*.	Sizes of frag.	# frag.	Sizes of frag.	# frag.	Sizes of frag.	# frag.	Sizes of frag.
DNP	7	2000, 100, 800, 700, 500, 400, 300	7	1000, 900**, 800, 700, 500, 400, 300	6	700, 600, 500,400, 300, 200	5	1000, 900, 800, 700, 600, 500, 400, 300, 200
RH	4	1000, 800, 500, 400	4	700, 500, 400, 300	3	500, 400, 300	8	2000, 1000, 900, 800, 700, 500, 300, 200
WN	4	2000, 800, 500, 300	3	600, 400, 300	3	800, 600, 500	7	3000, 2000, 800, 600, 500, 400, 300
SR	3	600**, 500, 400	4	500, 300, 200, 100***	5	700, 600, 400, 300, 200	5	1000, 700, 500***, 400, 300
ATB	4	900, 600, 400, 200	2	500, 300	2	500, 400	4	900, 700, 300, 200

* indicates DNA fragment i.e. the numbers of band sizes produced by each RAPD primer in each *P. orientalis* population.

**observed only in one *P. orientalis* specimen

*** observed only in three *P. orientalis* specimens

Table 5: Number of genotypes and DNA fragments produced by the RAPD-PCR in *Phlebotomus orientalis* individuals of each population.

P.orientalis Population	No of genotypes	No of fragments by RAPD (r469.1, r564.1, r751.2, r807.1)	No of fragments	No of fragments by RAPD (r469.1, r564.1, r751.2, r807.1)
DNP	62	18, 16, 9, 19	538	157, 112, 134, 135
RH	12	3, 3, 3, 3	674	141, 162, 148, 232
WN	9	2, 3, 2, 2	717	179, 84, 142, 312
SR	15	2, 6, 3, 4	302	69, 83, 66, 87
ATB	11	3, 2, 2, 4	112	38, 24, 22, 28
Total	109		2343	

3.4.2. Group-specific RAPD markers detected in *Phlebotomus orientalis* populations from different geographic areas in Sudan:

The four RAPD primers showed somewhat different degrees of heterozygosity among the tested *P. orientalis*. The results showed that r564.1 produced the highest numbers of group-specific bands (5 bands). However, *P. orientalis* populations of different geographic regions showed different numbers of RAPD specific markers by one primer or another; the exception was RH population which did not show specific markers. The WN population had specific bands (600, 800, 3000) by the three RAPD primers (r564.1, r751.2 and r807.1) compared to the other populations (table 6).

3.4.3. Variation at RAPD loci in *Phlebotomus orientalis* populations:

The frequencies of the alleles were found to be varied between individuals of the populations and between the populations as obtained for each RAPD primer (see table 7). The frequencies of alleles obtained by the four RAPD primers showed various level of genetic diversity among the individuals of all populations. However, the DNP population showed highest level of genetic diversity by any of the RAPD primers comparing to the other populations. The frequencies of alleles in RH, WN, SR and ATB populations were found to be varied between 0.00 and 1.00 by all RAPD primers. The frequencies of alleles detected in DNP populations were ranged from 0.00-0.76, 0.00-0.98, 0.00-0.5 and 0.00-0.96 by r469.1, r564.1, r751.2 and r807.1 respectively. Levels of genetic variation between populations were also indicated by percentage of polymorphic and expected heterogenicity as shown in table (7). The result showed the DNP had the highest percentage of polymorphic (P%

= 0.93) whereas, the WN had the lowest (P%= 0.41). Also, the expected heterogenicity (H_E) was found to be varied between the populations and, the values were found to be ranged from 0.13 to 0.17.

Table 6: Group specific bands amplified by the four RAPD primers obtained in each *Phlebotomus orientalis* **population from different geographic regions in Sudan.**

Population of P. orientalis	R469.1	r564.1	r751.2	R807.1
	Sizes of specific bands	Sizes of specific bands	Sizes of specific bands	Sizes of specific bands
DNP	700	1000, 900*, 800,	nd	nd
RH	nd	nd	nd	nd
WN	nd	600	800	3000
SR	nd	200, 100**	nd	nd
ATB	900, , 200	Nd	nd	nd

nd = not detected
* observed in only one *P. orientalis* specimen
** observed in only three *P. orientalis* specimens

Table 7: Allelic frequencies obtained by RAPD-PCR for the five populations of *P. orientalis* **from Sudan.**

	Locus	DNP	RH	WN	SR	ATB
	2000	0.04	0.00	0.58	0.00	0.0
	1000	0.36	0.30	0.00	0.00	0.00
	900	0.00	0.00	0.00	0.00	0.50
	800	0.66	0.52	1.00	0.00	0.00
	700	0.42	0.00	0.00	0.00	0.00
r469.1	600	0.00	0.00	0.00	0.00	1.00
	500	0.76	1.00	1.00	0.84	0.00
	400	0.72	1.00	0.00	1.00	1.00
	300	0.18	0.00	1.00	0.00	0.00
	200	0.00	0.00	0.00	0.00	0.33
	800	0.00	0.00	0.92	0.00	0.00
	700	0.30	0.18	0.00	0.06	0.00
	600	0.16	0.00	0.92	0.47	0.00
	500	0.48	1.00	0.00	0.00	1.00
r564.1	400	0.42	0.78	0.00	0.47	0.57
	300	0.98	1.00	1.00	0.06	0.00
	200	0.34	0.00	0.00	1.00	0.00
	3000	0.00	0.00	0.66	0.00	0.00
	2000	0.00	0.78	1.00	0.00	0.00
	1000	0.20	0.16	0.00	0.16	0.00
	900	0.16	0.62	0.00	0.00	0.00
	800	0.20	0.14	1.00	0.00	0.30
r751.2	700	0.48	1.00	0.00	0.55	0.23
	600	0.24	0.00	1.00	0.00	0.00
	500	0.50	1.00	1.00	0.10	0.00
	400	0.36	0.00	0.58	1.00	0.00
	300	0.42	0.78	1.00	1.00	1.00
	200	0.14	0.16	0.00	0.00	0.62
	1000	0.18	0.00	0.00	0.00	0.00
	900	0.02	0.00	0.00	0.00	0.00
	800	0.28	0.40	0.00	0.00	0.00
	700	0.18	0.00	0.00	0.19	0.00
	600	0.08	0.00	1.00	0.00	0.00
r801.1	500	0.96	1.00	0.00	1.00	0.85
	400	0.26	0.00	0.34	0.00	0.00
	300	0.28	0.92	0.34	0.68	1.00
	200	0.00	0.92	0.00	0.71	0.00
	100	0.00	0.00	0.00	0.10	0.00
P%		0.93	0.65	0.41	0.71	0.64
H_E		0.15	0.13	0.16	0.21	0.21

P% = Percentages of polymorphic loci
H_E = Expected heterozygosity

3.4.4. Genetic distance (Dice's similarity coefficient):

The results obtained by the Dice genetic distance (level of similarity matrix) showed different level of genetic variations among the individuals and between the populations. The genetic distances between individuals obtained by Dice coefficient were ranged from 0.45-0.735 in DNP, 0.432-0.735 in RH, 0.143-0.609 in WN, 0.412-0.652 in SR and from 0.143 to 0.581 in ATB. This result indicates lower genetic variation within the populations of DNP, RH and SR compared with that within the populations WN and ATB.

However, the genetic distance between the populations showed relatively high similarities between the populations of DNP and RH (0.735), DNP and SR (0.609) and RH and SR (0.649) (see table 8). The Dice coefficient showed low similarities between the populations of WN and ATB (0.143) and WN and SR (0.412) (see table 8).

The correlation between genetic (white) and geographic distance (gray) as indicated in Table 8 between the five populations of *P. orientalis* from Sudan were tested using a Mantel (1967) test, however, no association was observed between these populations (r=0.192) (see fig. 10).

Table 8: Geographic distance (gray) and genetic distance (white) between the populations of *P. orientalis* from different geographic regions in Sudan.

	DNP	RH	WN	SR	ATB
DNP		32.21	418.20	461.20	223.70
RH	0.735		411.80	451.50	191.60
WN	0.609	0.432		61.18	441.20
SR	0.652	0.649	0.412		457.50
ATB	0.450	0.581	0.143	0.500	

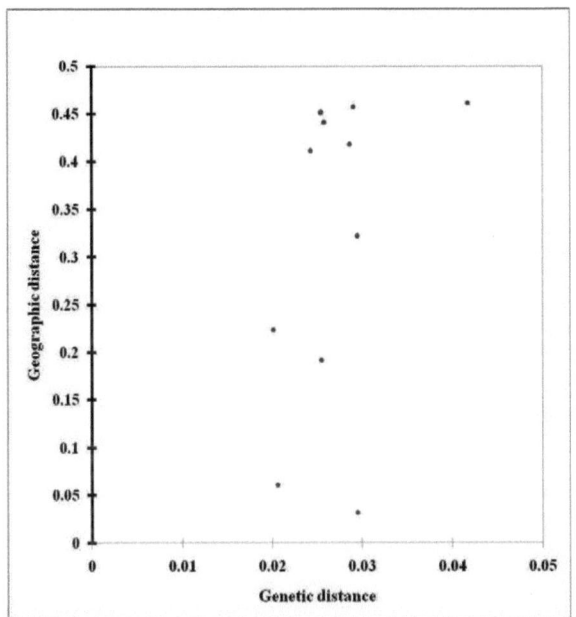

Figure 10: Correlation between genetic and geographic distance of populations of *Phlebotomus orientalis* **from five geographic regions in Sudan using a Mantel test with 1000 permutations (r = 0.476; P = 0.013).**

3.4.6. Neighbour-joining algorithm:

Phylogenetic trees obtained by each primer using RAPDistance software showed degree of variations. However, dendograms obtained by r469.1, r751.2 and r807.1 showed some degree of homogeneity between *P. orientalis* of each population which was indicated by the clustering of populations according to geographic regions whereas, the dendogram obtained by r564.1 showed high heterozygosity between these populations (see fig. 11, 12, 13 and 14). The dendogram which was obtained by r469.1 showed three main clusters DNP, WN and ATB whereas; RH and SR were clustered externally close to DNP (Fig11). The dendogram which was obtained by 564.1 showed two clusters DNP, RH except a few samples from the populations which grouped with SR and WN populations. Also ATB population and few individuals of SR population distributed between the populations of WN and RH (Fig 12). Primer r751.2 pointed three main clusters DNP, SR & WN, although three samples from RH, WN and few samples from ATB with SR. ATB populations were grouped externally closed to RH (Fig13). Primer r807.1 showed high similarity between RH and SR populations which grouped in one cluster and also between ATB and WN which clustered to each other, whereas DNP population except five samples formed a separate cluster (Fig. 14).

Neighbour-Joining dendogram generated from the genetic distance by all RAPD primers for all populations of *P. orientalis* is shown in figure 15. The hierarchical clustering analysis showed that *P. orientalis* clustered in five groups according to their geographic regions. However, the DNP and the RH populations showed high level of genetic similarity (0.74). The SR population was genetically more related to these two populations (0.65). In contrast, WN population was closer to the DNP, RH and SR populations (0.6) whereas, ATB was genetically less similar to all the populations (0.4).

PCA was performed to generalize geometric and the scores for the populations were plotted in figure 16 which encountered 73.06% of the total variance which agrees with that obtained by the dendograms in figure 15.

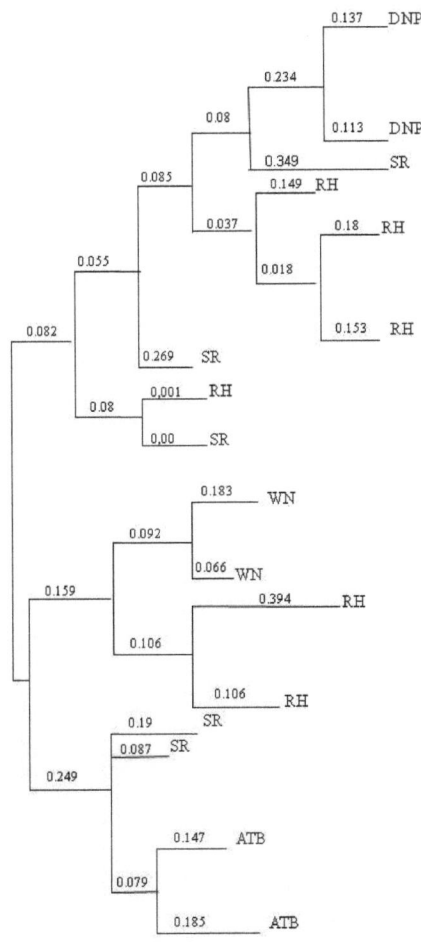

Figure 11: Neighbour-joining dendogram of Dice genetic distance (Nei's, 1978) obtained by r469.1 for *Phlebotomus orientalis* **populations from different geographic regions in Sudan.**

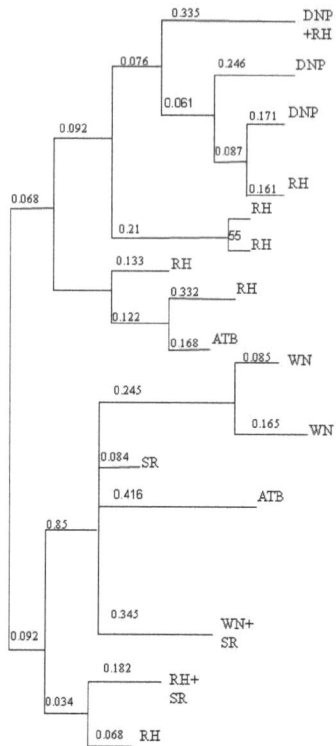

Figure 12: Neighbour-joining dendogram of Dice genetic distance (Nei's, 1978) obtained by r564.1 for *Phlebotomus orientalis* populations from different geographic regions in Sudan.

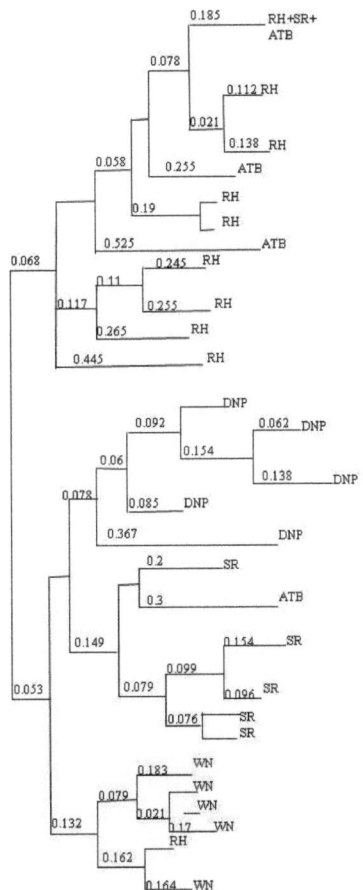

Figure 13: Neighbour-joining dendogram of Dice genetic distance (Nei's, 1978) obtained by r751.2 for *Phlebotomus orientalis* populations from different geographic regions in Sudan.

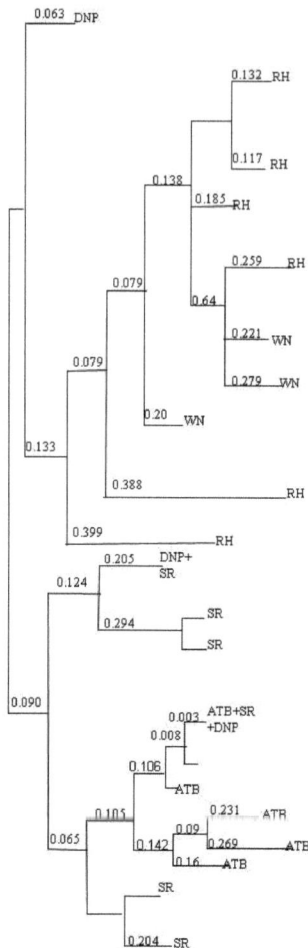

Figure 14: Neighbour-joining dendogram of Dice genetic distance (Nei's, 1978) obtained by r807.1 for *Phlebotomus orientalis* populations from different geographic regions in Sudan.

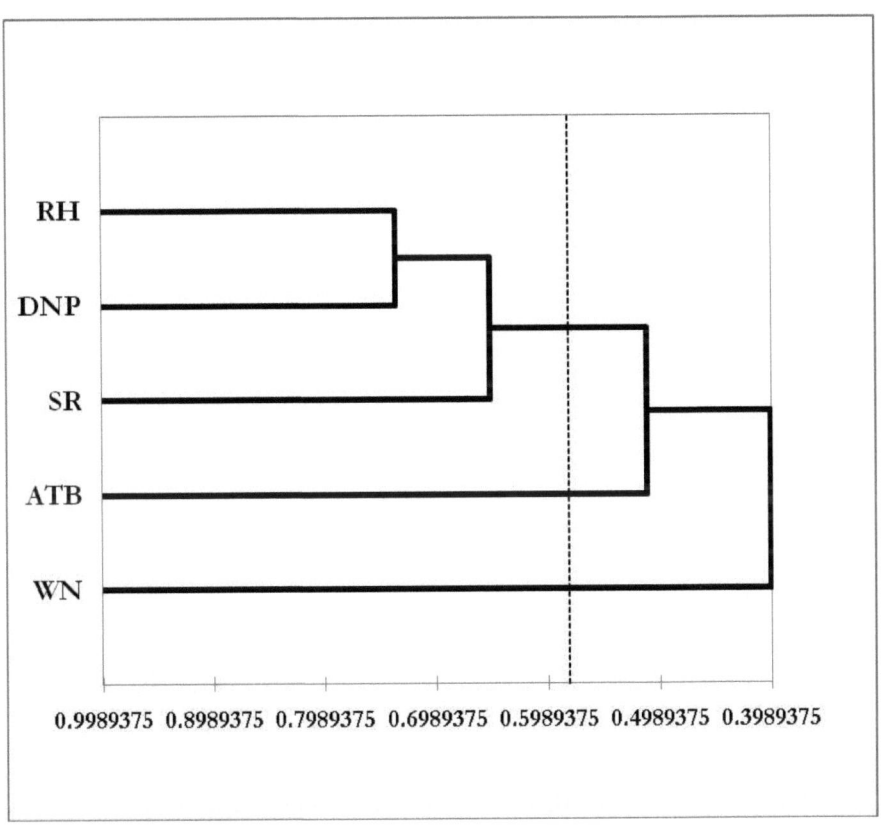

Figure 15: Neighbour-joining dendogram of Dice genetic distance (Nei's, 1978) obtained by all RAPD primers for *Phlebotomus orientalis* **populations from different geographic regions in Sudan.**

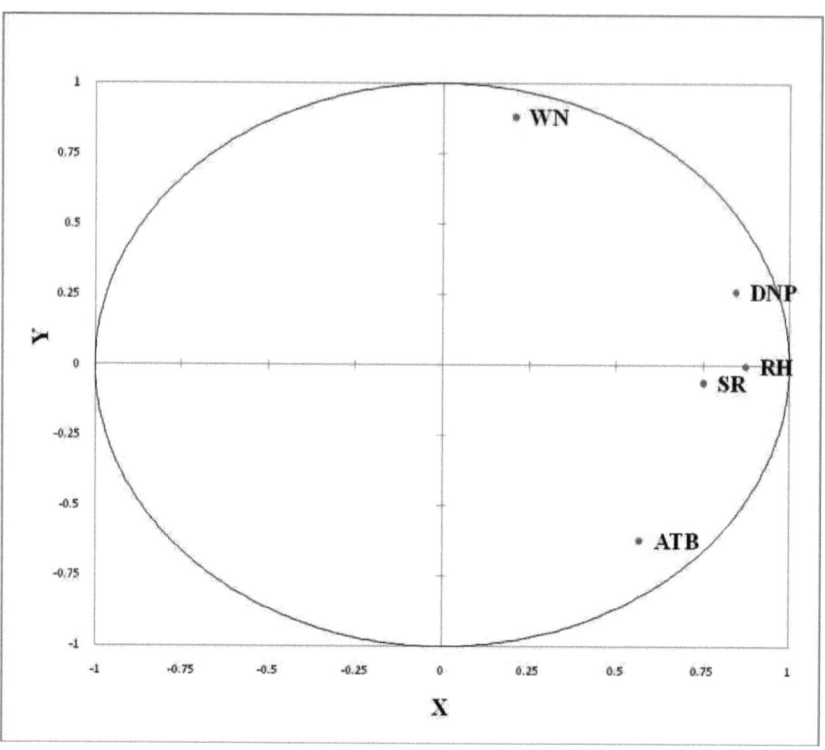

Figure 16: Principal coordinate analysis of populations of *Phlebotomus orientalis*
from different geographic regions in Sudan

3.5. Genetic structure of populations:

For determining if the genetic differences estimated from the sandflies are reflecting gene flow among insect population, fixation index (F_{ST}) was calculated using AMOVA approach (table 10). The AMOVA results showed a high proportion of total variation within populations (95.7%) but moderate proportion of variation would be attributed to either difference observed among groups within populations (4.3%) or among populations (1.6%). Patterns of genetic differentiation between populations of *P. orientalis* using measures of a fixation index F_{ST} showed a low genetic differentiation which indicates gene flow between these populations.

Table 9: **Molecular variance and fixation index of the populations and the individuals of** *Phlebotomus orientalis* **obtained by AMOVA test.**

Source of variations	df	Sum of squares	Variance components	% of variance	F_{ST}
Among populations	2	16.93	0.001	1.6	0.041
Among group within populations	2	19.74	0.02	4.3	0.042
Within individuals	2274	1059.8	0.45	95.7	0.002
Total	2278	1096.5	0.50		

CHAPTER FOUR

DISCUSSION

In Sudan, most of the studies on leishmaniasis vectors were directed towards ecology of sandflies, however, no single study has been conducted to elucidate the genetic variation and the population structure of *P. orientalis*. This is the first study to verify the genetic diversity and population structure of *P. orientalis* the vector of visceral leishmaniasis in Sudan. However, studies on the taxonomy of sandfly vectors remain a subject of interest because of their importance in transmission of leishmaniasis worldwide. Moreover, such studies will help our understanding the epidemiology pattern of the disease concerning the different forms of leishmaniasis due to same *Leishmania* strain. On the other hand this will help to set appropriate control strategy to eradicate the disease in each endemic area.

In this study, the result of sandflies composition recorded was in agreement with the general sandflies distribution in Sudan described by many authors (Lewis *et al.*, 1977; El-Sayed *et al.*, 1989; Elnaiem *et al.*, 1997; Lambert *et al.*, 2002; Hassan *et al.*, 2007). The authors recorded these species based on morphological adult characters, no entomological surveys were conducted in the White Nile area to identify sandfly species of the area, however, the sandflies fauna of Surogia village and eastern Sudan have been described (El-Sayed *et al.*, 1989; Elnaiem *et al.*, 1997; 1999; Lambert *et al.*, 2002; Hassan *et al.*, 2007).

Out of the sandflies collection, only two species are known to be important in view of transmission of leishmaniasis in Sudan, these are *P. orientalis* and *P. papatasi*. *Phlebotomus orientalis* is the proven vector of *L. donovani* the causative agent of VL in Sudan (Elnaiem *et al.*, 2001; Hassan *et al.*, 2004, 2008) whereas, *P. papatasi* has been incriminated as a vector of cutaneous leishmaniasis in arid areas of northern Sudan (El-Hassan and Zijlstra, 2001a). *Phlebotomus orientalis* has been found to occur in the savannah region dominated by *Acacia seyal-Balanities aegyptiaca* woodlands (Elnaiem *et al.*, 1998). However, in this study this species was collected from White Nile area and Surogia village in Khartoum State, a semi-desert region. This finding might indicate that existence of unexplored microclimates which provide an ideal determinant for *P. orientalis*.

Knowledge on sandfly biology is relatively limited in contrast to that of black flies and mosquitoes; this has had a negative impact on the development of a sound taxonomy for phlebotomine sandflies at the species level (Lane, 1986). Since the early days of the 20th century, most advances in the taxonomy of sandflies have followed from the studies on their role in the transmission of disease. Recent observations on geographic and local variation in behavior and morphology have led to speculations concerning presence of sibling species complexes in several of the sandfly vectors especially *Lutzomyia* sandflies. However, without a broadly based knowledge of the sandfly biology as well as its broad scale and microhabitat distribution, evolution of the variation in taxonomic and evolutionary terms has been problematic (Lane, 1986).

The uses of modern molecular techniques in different aspects of biology in general and entomology in particular is becoming the recent trend of pure an applied research. Therefore, this study was carried out to show the usefulness of the polymerase chain reaction (PCR) for identification of *P. orientalis* and the Random Amplified Polymorphic DNA (RAPD) to shed light on the population structure of *P. orientalis* in Sudan. For species identification, the primer pairs 2MMF, 2MMR were designed from a conserved region of mtDNA cytochrome b (Cyt b) and that because the information on sandflies genetics is very poor. Most of the studies were based on cytochrome b sequences of sand fly mitochondrial DNA (mtDNA) (Essenger *et al.,* 1997). Nevertheless, it is useful to study the evolutionary relationships - phylogeny - of organisms. Biologists can determine and then compare mtDNA sequences among different species and use the comparisons to build an evolutionary tree for the species examined.

In this study, *P. orientalis* was identified by the PCR amplification of the characteristic 675 bp DNA fragment specific for this species as suggested. The derived primers from mtDNA gene family provided high specificity (100%) to amplify DNA of *P. orientalis*, and failed to amplify the DNA extracted from other sandfly species. However, the morphological tools used for the differentiation of the species have been proved to be limited at the complex level. Hence, the advantage of using the species–specific PCR was its possibility to differentiate *P. orientalis* from other sandflies in the same area. Similarly, diagnostic primers were reported to be highly specific to amplify DNA of the targeted species and failed to differentiate between the subspecies as in case of *A. gambiae* complex (Collins & Paskwitz, 1996) or closely related species as in case of sandfly species, *P. papatasi* and *P. duboscqi* (Mukhopadhyay *et al.*, 2000).

In order to elucidate the level of genetic variation and the population structure of *P. orientalis* in Sudan, we employed RAPD analysis which has been described by Williams *et al.* (1990). RAPD analysis has been proved to be a powerful tool for systematic of species complexes and population studies even within well-established (species) that would otherwise go unnoticed (Wilkerson *et al.*, 1993). This technique has been expansively utilized in molecular parasitology as well as evolutionary studies for species identification and characterization of pathogens (Paskewtzi *et al.*, 1993). In molecular entomology, it proved to be of a great value in species identification, blood meal identification studies and insecticide resistance (Paskewitz

48

et al., 1993. Unlike other polymorphism assay no prior sequence knowledge is needed for the RAPD technique to be applied. Although RAPD polymorphism have been criticized for the lack of reproducibility among PCR runs made on the same organism, among individual members of species and among laboratories (Black, 1993), this is often inherent to the primers of the particular method. Many drawbacks of the RAPD-PCR technique can be overcome with a systematic approach and exhaustive testing (Grosberg *et al.,* 1996).

In this study, 200 specimens of *P. orientalis* from five different geographic regions in Sudan were analyzed by 30 RAPD primers. The primers were randomly selected based on G/C constituents. Only four RAPD primers amplification products were reproducible with characteristic common banding pattern. However, all visible bands were scored. Some strong bands were shared between the five areas, nevertheless, un similar RAPD banding patterns were also observed in few specimens from the same site, such bands were neglected as characterization criteria due to their in consistency. Individuals from the same area showed almost similar banding pattern with the four primers. The RAPD band patterns must be empirically determined to be reproducible before they have been used as markers (Williams *et al.,* 1990). It has been shown that even fairly strong RAPD bands will occasionally fail to be produced in particular amplification, thus inference based on the absence of a band should be made only after repeated reactions confirm the absence as reproducible (Williams *et al.,* 1990, Williams *et al.,* 1993; Wilkerson., *et al.,* 1993).

In the present study, baseline parameters of genetic variation were compared within and among population of *P. orientalis* from five widely separated locations in Sudan. The application of some population genetic principles in association with comparison of multi variable loci can give evidence in specific instances where the presence of a species complex is suspected. The genetic variation suggested by the four RAPD primers of *P. orientalis* apparently at odd with their morphological similarity. The isolates of *P. orientalis* in all five areas had similar morphological appearance. This may suggest a very close relationship and a relatively recent divergence. The possible explanation for this includes accelerated DNA sequence evolution and/or conservative morphological evolution.

The results obtained on total number of DNA fragments and the numbers of genotypes in *P. orientalis* were varied according to the origin of the populations. This number of fragments (2343) and genotypes (109) were sufficient to detect interpopulational differentiation (Wilkerson *et al.,* 1995). The presence of different genotypes of *P. orientalis* might indicate that these genotypes as being involved in specific transmission cycles of the parasites in each endemic area of VL (Cupeolillo *et al.,* 2003).

Morphological similarity implies close phylogenetic relationship and a recent speciation process, but it does not imply similarity of bionomics especially when dealing with sympatric taxa (Wright, 1943: White, 1982; Schliewen *et al.*, 1994). However, the overall amplification profiles generated by the four primers showed some level of genetic diversity among the *P. orientalis* from different geographic regions. Although they were generally quite similar, diagnostic DNA bands were detected among four of the five populations, the exception was Rahad population which did not give any specific bands. The result obtained on diagnostic RAPD markers may indicates the presence of subpopulations among the *P. orientalis* in Sudan. This finding might also suggest that each of the five geographic regions has it distinct *Leishmania* transmission pattern. In turn this finding may support the findings of Elamin *et al.* (2008) who found that *L. donovani* associated with cutaneous form of leishmaniasis in northern Sudan.

A high number of polymorphic RAPD-PCR loci are recommended for the identification of subspecies or lineages of organisms (Wilkerson *et al.*, 1995). However, no consensus in the literature indicates a recommended number of markers or numbers of primers to be used for the generation of reliable genetic profiles. In our case the four primers used generated about eight analyzable bands, with varied allele frequencies, varied percentages of polymorphic loci and varied level of heterozygosity within and among *P. orientalis* populations from different geographic regions.

In this study, the degree of genetic similarities between the five geographic populations of *P. orientalis* was initially estimated based on Dice coefficient (genetic distances). This method has been frequently used as the tool of choice in population biology, due to its statistical simplicity and for its suitability for analysis. Our results showed different levels of similarity between the *P. orientalis* populations from different geographic regions. These levels varied from high similarities as observed between populations DNP, RH and SR populations, moderate similarities as observed between WN, DNP and SR and low similarities as observed between WN population and the other populations (table 8). The high similarity between the DNP & the RH populations may be due to the similar climatic conditions and the geographic distance between the two areas (32 km). Clarck and Laingon (1993) have reported that local populations will not necessary show fixed differences. Although SR population occur is semi desert area and it differs from DNP and RH in the climatic conditions, also the geographic distance between these areas is high, the similarity between them was high.

In the present study, analysis of RAPD results using NJ tree, clustered the *P. orientalis* into five main groups according to their geographic regions with slight exceptions. This is an important since data for the other vectors such as *Lu.*

50

longipalpis incrimination methods including DNA props and isoenzyme studies have shown variation with geographic distribution (Arrivillaga *et al.* 2003; Balbino *et al.*, 2006). Similarly, The RAPD molecular approach has provided additional evidence to support the existence of bio-geographical populations of *Lu. witmani* in Brazil (Margonari *et al* 2004). Previous studies demonstrated that analysis of *Lu. Intermedia* populations using RAPD markers produced higher genetic variability than multilocus enzyme electrophoresis (Meneses *et al.*, 2005). Other studies on *Lu. intermedia* by De Souza Rosha *et al.* (2007) demonstrated that the number of polymorphic bands did not vary significantly according to the geographic origin of populations or microhabitat, and there was no RAPD marker specific to any one of the studied populations.

The absent of any significant correlation between genetic distance and geographical distance (Mantel test) suggest that the five regional populations are still isolated. It is unknown whether the barriers are prepostmating or environmental.

In this study, the degree of genetic structure between populations and among groups within populations was moderate which indicate that geographic separation of these populations was a recent event. The degree of variation within population is higher (95.7%) than between populations from localities separated by at least 32 km. This result corroborated other studies of population genetics of insect vectors, which demonstrated that the populations are homogeneous in radius of 20 km and that a certain degree of structuring is observed in populations encountered in zone of approximately 1 km (Dujardin *et al.*, 1987; Munstermann *et al.*, 1998; Marquenz *et al.*, 2001). However, two likely explanations for the low variation between the populations with in each area were suggested these explanations are: either some populations have become isolated but there has been no significant change in allele frequencies in any population or no population has become permanently isolated, and there continuing high gene flow between populations, over distance up to 500km. The latter seems more likely, although it could be difficult reconcile due to the supposedly 'weak flight capabilities' of sandflies (Cárdenas *et al.*, 2001). Weak flight has been used to explain the limited gene flow 'estimated from allozyme data' between Colombian populations of *Lu. longipalpis* (Morrison *et al.*,1993) or *Lu. shanonni* (Cardinas *et al.*, 2001; Mukhopadhayay *et al.*, 2001). However, it has been reported that the adults of *Phlebotomus* (*Larroussius*) species can travel up to 2.2km in as little 3 days during the Mediterranean summer (Killick-Kendrick *et al.*, 1984).

In conclusion, our results may suggest existence of subpopulations of *P. orientalis* in Sudan however; it is difficult to point out presence of species complex. However, before, to give a solid conclusion simultaneous analysis using different classes of molecular markers such as microsatellite is needed to provide answers towards the real taxonomic situation of the populations of *P. orientalis* in Sudan and to establish a true vector parasite relationship in the nature. Moreover, the genes amplified by RAPD primers in the present study might possibly be coding genes or even related to the coding genes associated with infectively or susceptibility of the vector to the insecticides. In this case, these differences observed between the different study sites become so important and hence further investigations would be followed aiming at mapping these regions and determining their functions. In this case, investigation would be sequencing of the diagnostic markers of *P. orientalis* of different populations which may provide more evidence on the level of variations among these populations. Thus good understanding of the vector population will lead to design effective control measures for reducing man-vector contact. Further *Leishmania* control programs should be based on genetic mapping of the vector.

REFERENCES

Abdalla, R. E., Ali, M., Wasfi, A. I. & El-Hassan, A. M. (1973). Cutaneous leishmaniasis in the Sudan. *Transactions of the Royal Society of Tropical Medicine & Hygiene*, **67**: 549-559.

Abonnenc, E. & Minter, D. M. (1965). Bilingual key for the identification of sandflies of the Ethiopian region. (French and English). *Cahiers ORSTOM, Sreie Entomologie Medicale*, **5**: 1-63.

Adamson, R. E., Ward, R. D., Felicingeli, M. D. & Maingon, R. (1993). The application of Random Amplified Polymorphic DNA for sandfly species identification. *Medical Veterinary & Entomology*, **7**: 203-207.

Alten, B., Caglar, S. S., Kaynas, S. & Simsek, F. M. (2003). Evaluation of protective efficacy of K-OTAB impregnated bednets for cutaneous leishmaniasis control in Southeast Anatolia-Turkey. *Journal of Vector Ecology*, **28** (1):53-64.

Altschul, S. F., Gish, W., Miller, W., Meyers, E. W. & Lipman, D. J. (1990). Basic local alignment search tool. *Journal of Molecular Biology*, **215(3)**: 403-410.

Aransay, A. M., Scoulia, E., Chaniotis, B . & Tselentis, Y. (1999). Typing of sandflies from Greece and Cyprus by DNA polymorphism of 18s rRNA gene. *Insect Molecular Biology*, **8**: 179-184.

Arrivillaga, J. C., Mutebi, J. P., Pianango, H., Norris, D., Alexander, B., Feliciangeli, M. D. & Lanzaro, G. C. (2003). The taxonomic status of genetically divergent populations of *Lutzomyia longipalpis* (Diptera: Psychodidea) based on the distribution of mitochondrial and isozyme variation. *Journal of Medical Entomology*, **40,** 615-627.

Arrivillaga, J. C., Norris, D. E., Feliciangeli, M. D. & Lanzaro, G. C. (2002). Phylogeography of the neotropical sand fly *Lutzomyia longipalpis* inferred from mitochondiral DNA sequences. *Infect Genet Evolution,* **2**: 83–95.

Ashford, R. W. (1996). Leishmaniasis reservoirs and their significance in control. *Clinical Dermatology,* **14**: 523-532

Ashford, R. W. & Bettini, S. (1987). Ecology and epidemiology: Old world In: The Leishmaniasis in Biology and Medicine Vol. 1. Edited by: Peters, W. and Killick-Kendrick, R. London: Academic Press, PP. 366-424.

Ashford, R. W., Seaman, J., Schorcher, J. & Pratlong, F. (1992). Epidemic visceral Leishmaniasis in southern Sudan: identify and systematic position of the parasite from patients and vectors. *Transactions of the Royal Society of Tropical Medicine & Hygiene,* **86**: 379-380.

Balbino, D. V., Coutinho-Aberu, I. V., Sonoda, I. V., Melo, A. M., de Andrade, P. P., de Castro, J. A., Rebelo, J. M., Carvalho, S. M. S., & Ohigao, M. R. (2006).

Genetic structure of the sandfly *Lutzomyia longipalpis* (Diptera: Psychodidae) from Brazilian north eastern region. *Acta Tropica,* **98**: 15-24.

Barroso, P. A., Marco, J. D., Kato, H., Tarama, R., Ruedai, P., Cajalo, S. P. & Basombri´o, M. (2007). The identification of sandfly species, from an area of Argentina with endemic leishmaniasis, by the PCR-based analysis of the gene coding for 18S ribosomal RNA *.Annals of Tropical Medicine & Parasitology,* **101**(3): 247–253.

Bartlett, J. & Stirling, D. (2003). A short history of the Polymerase Chain Reaction. In: methods in molecular biology. PCR Protocols. Vol 226. *2nd Edition(*edited by: Wasberg, R. B*.). Humana Press inc.* NewJersey, USA. PP3-6.

Benzie, J. A. H. & Wakeford, M. (1997). Genetic determination of sources of *Acanthaster planci* recrcitment. Technical Report N O.17 Townsville; CRC Reef Research Centre ltd, 31 pp.

Black, W.C. (1993). PCR with arbitrary primers: approach with care. *Insect Molecular Biology,* **2:** 1-16.

Cárdenas, E., Munstermann, L. E., Corredor, D., Martínez, O. & Ferro, C. (2001). Genetic variability among populations of *Lutzomyia shannoni (*Diptera: Psychodidae) from Colombia. *Memórias do Instituto Oswaldo Cruz,* **96**:189-196.

Clarck, A. G. & Laingon, C.M. (1993). Prospects for estimating nucleotide divergence with RAPD. *Molecular Biology & Evolulation,* **10**:1096-1111.

Collins, F. H., Mendez, M. A., Rasmussen, M. O., Mehaffey, P. C., Besansky, N. J. & Finnerty, V. (1987). A ribosomal RNA gene probe differentiated member of species of the *Anopheles gambiae complex. American Journal of Tropical Medicine & Hygiene,* **36**:37-41.

Collins, F. H. & Paskwitz, S. M. (1996). A review of the use of ribosomal DNA (rDNA) to differentiate among cryptic Anopheles species. *Insect Molecular Biology,* **5**:1-9.

Copeland, R. S., Koros, J., Ouko Taylor, M. K. A. & Roberts, C. R. (1992). Sensitivity of a ribosomal RNA gene for identification of life stages of *Anopheles arabiensis* and *Anopheles gambiae* (Diptera: Culicidae) using three storage methods. *Journal of Medical Entomology,* **29**:361-363.

Croen, D. G., Morrison, D. A., Ellis. J. T., (1997). Evolution of the genus *Leishmania* revealed by comparison of DNA and RNA polymerase gene sequences. *Molecular & Biochemical Parasitology,* **89**(2):149-159.

Crozier, R. H. & Crozier, Y. C. (1993). The mitochondrial genome of the honey bee *Apis millifera* : complete sequence and genome organization. *Genetics,* **133**(1):97.

Cupolillo, E., Brahim, L. R., Toaldo, C. B., Oliveira-Neto, M. P., Brito, M. E. F., Falqueto, A., Naiff, M. F. & Grimaldi Jr. G.(2003). Genetic polymorphism and molecular epidemiology of *Leishmania (Viannia) braziliensis* from different hosts and geographic areas in Brazil. *Journal of Clinical Microbiology,* **41**: 3126-3132.

Dasman, W. P. (1972). Development and Management of Dinder National Park and its Wildlife. *FAO, Rome, No. TA313, 61 PP.*

Davies, C. R., Llanos-Cuentas, E. A., Campos, P., Monge, J., Leon, E. & Canales, J. (2000). Spraying houses in the Peruvian Andes with lambda-cyhalothrin protects residents against cutaneous leishmaniasis. *Transaction of the Royal Society of Tropical Medicine & Hygiene,* **94**: 631–636.

Depaquite, J., Ferte, H., Leger, N., Lefranc, F., Alves Pire, C., Hanafi, H., Maroli, M., Morillas-Marquez, F., Rioux, J. A., Svobodova, M. & Volf, P. (2002). ITS 2 sequences heterogeneity in *Phlebotomus sergenti* and *Phlebotomus similes* (Diptera: Psychododae): possible sequences in their ability to transmit *Leishmania tropica. International Journal of Parasitology,* **32**: 1123-1131.

Deruere, J., El-Safi, S. H., Bucheton, B., Boni, M., Kheir, M. M., Davoust, B., Pratlong, F., Feugier, E., Lambert, M., Dessein, A. & Dedet, J. P. (2003). Visceral leishmaniasis in eastern Sudan: parasite identification in humans and dogs; host-parasite relationships. *Microbes Infections,* **5(12)**: 1103-1108.

Desjeux, P. (1991). Information on the epidemiological and control of leishmaniasis by country or territory. *WHO/ LEISH/ 91. 30.*

Desjeux, P. (1996). Leishmaniasis. Public health aspects and control. *Clinical Dermatology,* **14**: 417-423.

Desjeux, P. (2001). The increase in risk factors for leishmaniasis world wide. *Transactions of the Royal Society of Tropical Medicine & Hygiene,* **95**: 239-243.

De Souza Rosha, L., Falqueto A., Dos Santos, C. B., Gabriel, G. J. R., & Cupolillo, E. (2007). Genetic structure of *Lutzomyia (Nyssomia) intermedia* populations from two ecological regions in Brazil where transmission of *leishmania (vinnia) Braziliensis* reflects distinct eco-epidemiological feature. *American Journal of Tropical Medicine & Hygiene,* **76 (3)**: 559-565.

Diakou, A. & Dovas, C. I . (2001). Optimization of randomamplified polymorphic DNA producing amplicons up to 8500 bp and revealing intraspecific polymorphism in *Leishmania infantum* isolates. *Analytical Biochemistry,* **288**: 195–200.

Dias, E. S., Fortes-Dias, C. L., Stiteler, J. M., Perkins, P. V. & Lowyer, P. G. (1998). Random amplified polymorphic DNA (RAPD) analysis of *Lu .longipalpis* laboratory populations. *Revista do Instituto de Medicina Tropical de São Paulo,* **40**: 49-53.

Dujardin, J. P., Tibayrene, M., Venegas, E., Maldonado, L., Desjeux. P. & Ayala. F. J. (1987). Isozyme evidence of lack of speciation between wild and domestic
55

Triatoma infestans (Heteropetra, Reduviidae) in Bolivia. *Journal of Medical Entomology*, **24**: 40-45.

El-Hassan, A. M., & Zijlstra, E. E. (2001a). Leishmaniasis in Sudan. Cutaneous leishmaniasis. *Transaction of the Royal Society of Tropical Medicine & Hygiene*, **95** (Suppl 1): S1-17.

El-Hassan A. M & Zijlstra, E. E,. (2001b). Leishmaniasis in Sudan. Mucosal leishmaniasis. *Transaction of the Royal Society of Tropical Medicine & Hygiene*, **95** (Suppl 1): S19-26.

El-Hassan, A. M., Zijstra, E. E., Ismail, A. & Ghalib, H. W. (1995). Recent observations on the epidemiology of kala-azar in eastern and central states of the Sudan. *Tropical Geographical Medicine*, **47**:151-156.

Elnaiem, D. A., Conners, S., Thmoson, M., Hassan, M. M., Hassan, K. H., Aboud, M. A. & Ashford, R. W. (1998b). Environmental determinants of the distribution of *Phlebotomus orientalis* in Sudan. *Annals of Tropical Medicine & Parasitology*, **92**: 2877-2887.

Elnaiem, D. A., Hassan, K. H., & Ward, R. D. (1997). Phlebotomine sandflies in a focus of visceral leishmaniasis in a border area of eastern Sudan. *Annals of Tropical Medicine & Parasitology*, **91**: 307-318.

Elnaiem, D. A., Hassan, K. H., Ward, R. D., Miles, M. A. & Frame, I. A. (1998a). Infection rates of *Leishmania donovani* in *Phlebotomus orientalis* from visceral Leishmaniasis focus eastern Sudan. *Annals of Tropical Medicine & Parasitology*, **92**: 229-232.

Elnaiem, D. A., Hassan, M. M., Maingon, R., Nureldan, G. H., Mekkawi, A. M., Miles, M., & Ward, R. D. (2001). The Egyptian mangoose, *Herpests inchneumon*, is a possible reservoir host of visceral leishmaniasis in eastern Sudan. *Parasitology*, **122(5)**: 531-536.

Elnaiem, D. A., Meneses, C., Slotman, M. & Lanzaro, G. C. (2005). Genetic variation in the sand fly salivary protein, SP-15, a potential vaccine candidate against *Leishmania major. Insect Molecular Biology*, **14**:145–150.

Elnaiem, D. A., Schorscher, J., Bendall, A., Obsomer, V., Osman, M. E., Mekkawi, A. M., Connor, S. J., Ashford, R. W. & Thomson, M. C. (2003). Risk mapping of visceral leishmaniasis: the role of local variation in rainfall and altitude on the presence and incidence of kala-azar in eastern Sudan. *American Journal of Tropical Medicine & Hygiene*, **68(1)**: 10-17.

El-Safi, S. H. & Peters, W. (1991). Studies on the leishmaniasis in the Sudan. Endemic of cutaneous leishmaniasis. *Transaction of Royal Society of the Tropical Medicine & Hygiene*, **85**: 44-47.

El-Safi, S. M., Peters, W., El Toam, B., El-Kadaro, A. & Evans, D. A. (1991). Studies on the leishmaniasis in Sudan. 2. Clinical and parasitological studies on cutaneous leishmaniasis. *Transactions of the Royal Society of Tropical Medicine & Hygiene*, **85:** 457-464.

El-Sayed, S.M., El-Raaba, F.M. & Abd el Nur, O. (1991). Daily and seasonal activities of sandflies from Surogia village, Khartoum, Sudan. *Parasitologia,* **33(suppl):** 205-15.

El-Sayed, S. M., Hemingway, J., and Lane, R. P. (1989). Susceptibility baselines for DDT metabolism and related enzyme systems in the sandfly *Phlebotomus papatasi* (Scopali) (Diptera: Psychodidae). *Bulletin of Entomological Research,* **79:** 679-684.

Ernest, D. & El Wasila, M. (1985). Recent situation of Dinder National Park in the Sudan in "Proceeding on Wildlife Conservation and Management in Sudan", Khartoum" 16-21 March 1985. *(eds, Ernest D.). Buch und offsetduckereia Guner Stubbemann Gum, Humburg, 1991.*

Esseghir, S., Ready, P.D. & Ben-Ismail, R. (2000). Speciation of *Phlebotomus* sandflies of the subgenus *Larroussius* coincided with the late Miocene-Pliocene aridification of the Mediterranean subregion. *Biological. Journal of the Linnean Society of London,* **70(2):** 189-219.

Esseghir, S., Ready, P. D., Killick-Kendrick, R. & Ben-Ismail, R. (1997). Mitochondrial haplotypes and phylogeography of *Phlebotomus* vectors of *Leishmania major. Insect Molecular Biology,* **6:** 211-225.

Excoffier, L., Laval, G. & Schneider, S. (2006). Arlequin , version 3.1: An Integrated Software Package for Population Genetics Data Analysis. Switzerland: omputational and Molecular Population Genetics Laboratory, University of Berne.

Felserstein, J. (1981). Evolutionary trees from DNA sequences: a maximum likelihood approach. *Journal of Molecular Evolution,* **17(6):** 368-376.

Fryauff. D & hanafi, H. (1991). Demonstration of hybridization between *Phlebotomus papatasi* (Scopoli) and *Phlebotomus bergeroti* Parrot. *Parassitologia,* **33(Suppl. 1):** 237-243.

Gale, K. R. & Crampton, J. M. (1987). DNA probes for species identification of mosquitoes in the *Anopheles gambiae* complex. *Medical and Veterinary Entomology,* **1:** 127-136.

Gebre-Michael, T., Malone, J. B., Balkew, M., Ali, A. & Berhe, N. (2004). Mapping the potential distribution of *Phlebotomus martini* and *P. orientalis* (Diptera: Psychodidae), vectors of kala-azar in East Africa by use of geographic information systems. *Acta Tropica,* **90:** 73–86.

Ghosh, K. N., Mukhopadhyay, J., Guzman, H., Tesh, R. B. & Munstermann, L. E. (1999). Interspecific hybridization and genetic variability in strains of phlebtomine sandflies. *Medical Veterinary & Entomology,* **13**: 78-88.

Grosberg , R. K., Levitan, D. R. & Cameron, B. B. (1996). Characterization of genetic structure and genealogies using RAPD-PCR markers: A random primer for the novice and nervous. In: Ferraris, J. D., Palumbi, S. R. (Eds), molecular zoology: Advances, Strategies, and Protocols. Wiley-Liss, New York, PP. 67-100.

Hartl, D. L. (1988). A primer of population genetics. *Sinauer. Associate, Senderland, MA.*

Hassan, M. M., Elamin, E. M. & Mukhtar, M. M. (2008). Isolation and identification of *Leishmania donovani* from *Phlebtomus orientalis,* in an area of eastern Sudan with endemic visceral lieshmaniasis. *Annals of Tropical Medicine &Parasitology,* **102 (6)** 553-555.

Hassan, M. M., El-raba'a, F. M. A., Ward, R. D., Maingon, R. D. C. & Elnaiem, D. A. (2004). Detection of high rates of in-village transmission of *Leishmania donovani* in eastern Sudan. *Acta Tropica,* **92 (1)**: 77-82.

Hassan, M. M., Osman, O. F., El-Raba'a, F. M. A., Schalling, H. D. F. H. & Elnaiem, D. A. (2009). Role of the domestic dog as a reservoir host of *Leishmania donovani* in eastern Sudan. *Parasites and Vectors,* **2**: 26.

Hassan, M. M., Widaa, S. O., Ibrahim, M. A., AbuShara, R., Osman, M. O., Numairy, M. S. M. & Khider, E. M. (2007). Studies on ecology of sand flies in Sudan: the First Records of *Phlebotomus orientalis* and *Phlebotomus rodhaini* (Diptera: psychodidae) in Northern Sudan. *Annals of Tropical Medicine & Parasitology,* **101 (7)**: 653-655.

Hayashi, K & Yandle, D. W. (1993). Howsensitive is PCR-SSCP. *Human Mutation,* **215**:338-334.

Hiss, R. H., Norris, D. E., Dietrich, C. H., Whitcomb, R. F., West, D. F. & Black, W. C. (1994). Molecular taxonomy using single- strand conformation polymorphism (SSCP) analysis of mitochondrial ribosomal DNA genes. *Insect Molecular Biology,* **3(3)**: 171-182.

Hoogstraal, H. & Heyneman, D. (1969). Leishmaniasis in Sudan Republic. 30. Final epidemiological Report. *American Journal of Tropical Medicine & Hygiene,* **18**: 1091-1210.

Ishikawa, E. W. Y., Ready, P. D., Souza, A. A., Day, J. C., Rangel, E. F., Davies, C. R. & Shaw, J. J. (1999). A mitochondrial DNA phylogeny indicates close relationships between populations of *Lutzomyia whatmani* (Diptera; Psychodidae; Phlebotominae) from the rain-forest regions of Amazon and Northeast Brazil. *Memórias do Instituto Oswaldo Cruz,* **94**: 339-345.

Jeanmougin, F., Thompson, J. D., Gouy, M., Higgins, D. G. & Giblbson, T. J. (1998). Multiple sequence alignment with Clustal X. *Trends in Biochemical Science,* **23(10)**:403-405.

Jiménez, M. E., Bello, F. J., Ferro, C. & Cárdenas, E. (2001). Brain cell karyotype of the phlebotomine sand fly *Lutzomyia shannoni* (Diptera: Psychodidae*). Memórias do Instituto Oswaldo Cruz,* **96(3):** 379-380.

Khamesipour, A., Dowlati, Y., Asilian, A., Hashemi-Fesharki, R., Javadi,A., Noazin, S. & Modabber, F. (2005). Leishmanization: use of an old method for evaluation of candidate vaccines against leishmaniasis. *Vaccine,* **23**, 3642–3648.

Killick-Kendrick R. (1990). Phlebotomine vectors of leishmaniases*: a review of Medical & Veterinary Entomology,* **4**: 1-24.

Killick-Kendrick, R. & Killick-Kendrick, M. (1987). Honey-drew of aphids as a source of sugar for *Phlebotomus ariasi. Medical & Veterinary Entomology,* **1**: 297-302.

Killick-Kendrick, R., Rioux, J. A., Bailly, M., Guy, M. W., Wilkes, T. J., Guy, F. M., Davidson, I., Knechtli, R., Ward, R. D., Guilvard, E., Perieres, J. & Dubois, H. (1984). Ecology of leishmaniasis in the south of France. 20. Dispersal of *Phlebotomus ariasi Tonnoir*, 1921 as a factor in the spread of visceral leishmaniasis in the Cévennes. *Annales de Parasitologie Humaine et Comparee (Paris),* **59**: 555-572.

Kirk, R. & Lewis, D. J. (1947). Studies in leishmaniasis in the Anglo-Egyptian Sudan. IX. Further Observations on the Sandflies *(Phlebotomus)* of the Sudan. *Transaction of the Royal Society of Medicine & Hygiene,* **40**: 869-888.

Kirk R & Lewis D. J. (1951). The phlebotomine of the Ethiopian region. *Transactions of Royal Entomological Society of London,* **102**: 383-510.

Kirk, R. & Lewis, D. J. (1955). Studies in leishmaniasis in the anglo-Egyptian Sudan x1.Phlebtomines in relation to leishmaniasis in Sudan. *Translations of the Royal Society of Tropical Medicine and Hygiene,* **49**:229-240.

Klaus, S. & Frankenburg, S. (1999). Cutaneous leishmaniasis in the Middle East. *Clinical Dermatology,* **17**: 137-141.

Lainson, R. (1988). Ecological interaction in the transmission of the leishmaniasis. *Philosophical Transactions of the Royal Society of London,* **321**: 389-404.

Lambert, M., Dereure, J., El-Safi. S., Bucheton, B., Dessien, A., Boni, M., Feugier, E. & Dedet, J.P.(2002). The sandfly fauna in the visceral-leishmaniasis focus of Gedaref, in the Atbara-River area of eastern Sudan. *Annals of Tropical Medicine and Parasitology*, 69 (6):631.

Lane, R. P. (1986). Recent advances in the systematics phlebotomine sand flies. *Insect Science Application,* **7**: 225-230.

Lane, R. P. (1993). Sandflies (Phlebotomine). In: Medical Insects and Arachnids. (eds, Lane R. P. & Crosskey, R. W). Chapman & Hall, London. pp. 78-119 .

Lanzaro, G. C., Lopes, A. H. C. S., Ribeiro, J. M. C., Shoemaker, C. B., Warburg, A., Soares, M. and Titus, R. G.(1999). Variation in the salivary peptide, maxadilan, from species in the *Lutzomyia longipalpis* complex. *Insect Molecular Biology*, **8**:1-9.

Lanzaro, G. C., Ostrovska, K., Herrero, M. V., Lawyer, P. G. & Warbur, G A. (1993). *Lutzomyia longipalpis* is a species complex: genetic divergence and interspecific hybrid sterility among three populations. *American Journal of Tropical Medicine & Hygiene*, **48**: 839-84.

Lanzaro, G. C. & Warbrug, A. (1995). Genetic variability in phlebtomine sandflies: possible implications for leishmaniasis epidemiology. *Parasitolgia Today*, 11: 151-154.

Le Blancq, S. M. & Peters, W. (1986). Leishmaniasis in the Old World: 4. The distribution of *Leishmania donovani* senso lato zymodemes. *Transactions of the Royal Society of Tropical Medicine and Hygiene*, **80 (3):** 367-377.

León, M.A., Shaw, J. J., & Tapia, F. J. (1996) A guide for the cutaneous leishmaniasis connoisseur. In Molecular and Immune Mechanisms in the Pathogenesis of Cutaneous Leishmaniasis (Tapia, F. J., Cáceres-Dittmar, G., and Sánchez, M. A., eds.), Intelligence Unit Series. R. G. Landes Bioscience Publishers, TX, pp. 1–23.

Lewis, D. J. (1973). Phlebotomidae and Psychodidae (sandflies and moth-flies). In: Insect and other Arthropods of Medical Importance (eds. Smith K. G. V.). The Trustees of the British Museum (Natural History), London. pp. 155-180.

Lewis, D. J., Young, D. G., Fairchild, G. B. & Minter, D. M. (1977). Proposal for stable classification of Phlebotomine sandflies *(Diptera: Psychodidae)*. *Systematic Entomology*, **2**: 319-332.

Madulo-Leblond, G., Killick-Kendrick, R., Killick-Kendrick, M. & Pesson, B. (1991). Comparison entre *Phlebotomus duboscqi* Nevou-Lemaire, 1906 et *Phlebotomus papatasi* (Scopoli, 1786): etudes morphologique et isoenzymatique. *Parasitologia*, **33 (supl. 1)**: 387-391.

Maingon, R. D. C., Ward, R. D., Hamilton, J. G. C., Noyes, H. A., Souza, N., Kem. S. J. & Watt, P. C. (2003). Genetic identification of two sibling species of *Lutzomyia longipalpis* (Diptera: Psychodidae) that produce distinct male sex pheromones in Sobral, Ceara state, Brazil. *Molecular Ecology*, **12**: 1879-1894.

Mancini, P., Martin-Sanchez, P., Maroli, M. & Gramiccia, M. (1997). Phylogenetic analysis of four *Phlebotomus* species (Diptera: Psychodidae*)* by ITS2 rDNA sequences. *Acta Parasitologica*, **21**: 166-167.

Mangabeira-Filho, O. (1969). Sôbre a sistemática e biologia dos *Phlebotomus* do Ceará. *Revista Brasileira de Malariologia e Doencas Tropicais,* **21:** 3-26.

Mantel, N. (1967). The detection of disease clustering and generalized regression approach. *Cancer Research,* **27**: 209-220.

Margonari, C. S., Dias-Fortes, C. L. & Dias, E. S. (1998). Molecular studies of geographical populations of *Lutzomyia witmani* (Diptera, Psychodidae. Phlebtominae) by RAPD-PCR. *Memórias do Instituto Oswaldo,* **93(sup. 2)**: 336-337

Margonari, C. S., Dias-Fortes, C. L. & Dias, E. S. (2004). Genetic variability in geographical populations of *Lutzomyia whitmani* elucidated by RAPD-PCR. *Journal of Medical Entomology,* **41**: 187-192..

Marquenz , L., M., Lambo, M., Rinaldi, M. & Lau, P. (2001). Gene flow between natural and domestic populations of *Lutzomyia longipalpis* (Diptera: Psychodidae) in restricted focus of American visceral leishmaniasis in Valenzuela. *Journal of Medical Entomology,* **38**:12-16.

Martin-sauchez, J., Gramiccia, M., Pesson, B., & Morillas-Marquez, F. (2000). Genetic polymorphism in sympatric species of the genus *Phlebtomus,* with special reference to *Phlebotomus perniciosus* and *Phlebotomus longipalpis* (Diptera: phlebtomidae). *Parasite,* **7**: 247-254.

Meneses, C.R., Cupolillo, E., Monteiro, F. & Rangel, E.F. (2005). Microgeographical variation among male populations of the sandfly, *Lutzonyia (Nyssomyia) intermedia,* from an endemic area of American cutaneous leishmaniasis in the state of Rio de Janeiro, Brazil. *Medical and Veterinary Entomology,* **19**:38-47.

Miles, S. J. (1978). Enzyme variation in the *Anopheles gambiae* Giles group of species (Diptera: Culcidae). *Bulletin of Entomological Research,* **68**:85-96.

Miles, M. A. & Ward, R. D. (1978). Preliminary isoenzyme studies on phlebtomine sand flies (Diptera: psychodidae). *Annals of Tropical Medicine and Parasitology,***72**: 398-400.

Minter, D. M., Wijers, D. J. B., Heisch, R. B. & Manson-Bahr, P. E. C. (1962). *Phlebotomus martini,* a probable vector of kala-azar in Kenya. *British Medical Journal,* **ii**: 835.

Morrison, A. C., Ferro, C., Morales, A., Tesh, R. B. & Wilson, M. L. (1993). Dispersal of the sandfly *Lutzomyia longipalpis* (Diptera : Psychodidae) at an endemic focus of visceral leishmaniasis in Colombia. *Journal of Medical Entomology,* **30**:427-435.

Mortz , G. & Hillis, D. M. (1987). *Molecular systematics. Sinauer.*

Muccio, T. D., Marinucci, M., Frusteri, L., Maroli, M., Pesson, B. & Gramiccia, M. (2000). Phylogenetic analysis of *Phlebotomus* species belonging to the subgenus

Larroussius (Diptera, Psychodidae) by ITS2 rDNA sequences. *Insect Biochemistry & Molecular Biology,* **30**: 387–393.

Mukhopadhlyay, J., Ghosh, K. & Braig H. R. (2000). Identification of cutaneous leishmaniasis vectors. *Phlebotomus papatasi* and *P. duboscqi* using random amplified polymorphic DNA. *Acta Tropica,* 76: 277-283.

Mukhopadhayay, J., Ghosh, K., Ferro, C. & Munstermann, L. E. (2001). Distribution of *Phlebotomus* sandflies genotypes (*Lutzomyia shanonni* Diptera: Psychodidae) across a highly hetrogeneous landscape. *Journal of Medical Entomology,* **38**: 260-267.

Mukhopadhyay, J., Ghosh, K., Rangel, E.F. & Munstermann, L.L., (1998). Genetic variability in biochemical characters of Brazilian field populations of the *Leishmania* vector, *Lutzomyia longipalpis* (Diptera:Psychodidae). *American Journal of Tropical Medicine and Hygiene,* **59**: 893-901.

Mullis, K. B. & Faloona, F.A. (1987) Specific synthesis of DNA in vitro via polymerase catalyzed chain reaction. *Methods Enzymology,* **155**:335-350.

Munsterman, L. E., Morrison, A. C., Ferro, C., Pardo, R. & Torres, M. (1998). Genetic structure of local population of *Lutzomyia longipalpis* (Diptera: Psychodidae) in central Colombia. *Journal of Medical Entomology,* **35**:82-89.

Murphy, R. W., Sites, JR. JW., Buth, D. G. & Haufler, C. N. (1990). Protein 1: isozyme electrophoresis In: molecular systematic. (eds Hillis, D. M. and Mortz, G.) Sinauer Associates, Sunder land MA. pp 45-126.

Mutebi, J., Alexander, B., Sherlock, I., Wellington, J., Souza, A. A., Shaw, J., Rangel, E. F. & Lanzaro, G.C. (1999). Breeding structure of the sandfly *Lutzomyia longipalpis (Lutz* and *Neiva)* in Brazil. *American Journal of Tropical Medicine & Huygens,* **61**: 149-157.

Nei, M. (1978). Estimate of average hetrozygosity and genetic distance from a small number of individuals. *Genetics,* **89**: 583-590.

Orita, M., Iwahana, H., Kanazawa, H., Hayashi, K & Sekia,T. (1989). Detection of Polymorphisms of Human DNA by Gel Electrophoresis as SSCPs. *Proceedings of the National Academy of Sciences of the United States of America,* **86**: 2766-2670.

Oskam, L., Pratlong, F., Zijlstra, E. E., Kroon, C. C. M., Dedet, J. P., Kager, P. A., Schijnian, G., Ghalib, H. W., El-Hassan, A. M. & Meredith, S. E. 0. (1998). Biochemical and molecular characterization of *Leishmania* parasites isolated from an endemic focus in eastern Sudan. *Transactions of the Royal Society of Tropical Medicine and Hygiene,* **92**: 120- 122.

Osman, O. F., Kager, P. A. & Oskam, L. (2000). Leishmaniasis in the Sudan: a literature review with emphasis on clinical aspects. *Tropical Medicine & International Health,* **5(8)**: 553-562.

Pape, T. (1992). Cladistic analysis of mosquitoes chromosome data. *Mosquito Systematics,* **24**: 1-11.

Paskewitz, S. M., Coetzze, M. & Hunt, R. H.(1993). Evolution of the polymerase chain reaction method for identifying members of the *Anopheles gambiae* (Diptera: Culicidae) complex in South Africa. *Journal of Medical Entomology,* **30(5)**: 953-957.

Paskewitz, S. M. & Collins, F. H. (1990). Use of polymerase chain reaction to identify mosquito species of the *Anopheles gambiae* complex. *Medicine & Veterinary Entomology,* **41**: 367-373.

Quate, L. W. (1964). Leishmaniasis in Sudan Republic. 19. *Phlebotomus* sandflies of the Paloich area in the Sudan (Diptera; Psychodidae). *Journal of Medical Entomology,* **1**: 213-268.

Ready, P. D. (1979). Factors affecting egg population of laboratory bred *Lutzomyia longipalpis* (Diptera; Psychodidae). *Journal of Medical Entomology,* **16**: 413-423.

Russell, R., Iribar, M. P., Lambson, B., Brewster, S., Blackwell, J. M., Dye, C. & Ajioka, J. W. (1999). Intra and inter-species microsatellite variation in the *Leishmania* subgenus *Viannia. Molecular Biochemistry & Parasitology,* **103**:71-77.

Saitou, N., & Nei. M. (1987). The neighbour-joining method, a new method for reconstructing phylogenetic trees. *Molecular Biology & Evolution, 4,406-425.*

Sambrook, J., Fritsch, E. F. & Maniatis, T. (1989). Molecular Cloning: a laboratory Manual, 2nd edition. Cold Spring Harbor laboratory press, Cold Spring Harbor, NewYork.

Schlein, Y,. Jacobson, R.L. & Muller, G. (2001). Sand fly feeding on noxious plants: a potential method for the control of Leishmaniasis. *American Journal of Tropical Medicine & Hygiene,* **65**:300-303.

Schliewen, U. K., Tautz, D. & Paabo, S. (1994). Sympatric speciation suggested by monophyly of crater lack cichlids. *Nature,* **368**: 629-632.

Schorscher, J. A. & Goris, M. (1992). Incrimination of *Phlebotomus (Larroussius) orientalis* as a vector of visceral Leishmaniasis in Western Upper Nile Province, southern Sudan. *Transactions of the Royal Society of Tropical Medicine & Hygiene,* **86**: 622-623.

Seaman, J., Mercer, A. J. & Sondrop, G. (1996). The epidemic of visceral leishmaniasis in Western Upper Nile, southern Sudan: Course of impact from 1984 to 1994. *International Journal of Epidemiology,* **25(4)**: 862-871.

Siaki, R. K., Gelfand, D. H., Stoffel, S., Scharf, S. J., Higuchi, R., Horn, G. T., Mullis, K. B. & Ehrilch, A. (1988). Primer directed enzymatic amplification of DNA with thermostable DNA polymerase. *Science,* **239**: 487-491.

Subbaro, S.K. (1997). Anopheline species complexes and malaria control. *Indian Journal of Medical Research,* **106**: 164-73.

Tauil, P. L. (2006). Perspectivas de controle de doenças transmitidas por vetores no Brasil. *Revista da Sociedade Brasileira de Medicina Tropical,* **39**: 275-277.

Testa, J. M., Montoya-Lerma, J., Cadena, H., Oviedo, M. & Ready P. D. (2002). Molecular identification of vectors of Leishmania in Colombia: Mitochondrial introgression in the *Lutzomyia townsendi* series. *Acta Tropica,* **84**: 205-218.

Thomson, M. C., Elnaiem, D. A., Ashford, R. W. & Connor, S. J. (1999). Towards a kala-azar risk map for Sudan: mapping the potential distribution of *Phlebotomus orientalis* using digital data of environmental variables. *Tropical Medicine & International Health,* **4(2)**:105-13.

Torgerson, D. G., Margarita, L., Velazquez,Y. & Patrick T. K. W.(2003). Genetic relationships among some species groups within the genus *Lutzomyia* (Diptera: Psychodidae). *American Journal of Tropical Medicine & Hygiene,* **69(5)**, 484-493.

Valenzuela, J. G., Belkaid , Y., Garfild, M. K., Mendoz, S., Kamhawi, S., Rowton, E., Sacks, D. L. & Ribeiro, J. M. C. (2001). Towards a defined anti-*Leishmania* vaccine targeting vector antigens: characterization of a protective salivary protein. *Journal of Experimental Medicine,* **194**:331-342.

Warburg, A., Saravia, E., Lanzaro, G. C., Titus, R. & Neva, F. (1994). Saliva of *Lutzomyia longipalpis* sibling species differs in its composition and propensity to enhance leishmaniasis. *Philosophical Transactions of the Royal Society London,* **345**: 223-230.

Ward, R. D. (1985). Vector biology and control In: Human Parasitic Diseases. *Vol. 1. Leishmaniasis (eds. Cang K. P. & Bray R. S.). Elsevier publisher, Amsterdam, New York, Oxford,* Pp. 199-212.

Ward, R. D. (1989). Some aspects of the biology of phlebotomine sandfly vectors. In Advances in Disease Vector Research, ed. Harris, K F. Springer-Verlag, New York, pp91-126.

Ward, R.D & Morton, I. E. (1991). Pheromones in mate choice and sexual isolation between siblings of *Lutzomyia longipalpis* (Diptera: Psychodidae). *Parassitologia* 33, Supplement 1, 527-533.

Ward, R.D., Pastiur, N. & Rioux, J.A. (1981). Electrophoretic studies on genetic polymorphism and differentiation of phlebotomine sand flies in France and Tunisia. *Annals of Tropical Medicine & Parasite,* **56**:19-23.

Ward, R. D., Phillips, A., Burnet, B. & Marcondes, C. B. (1988). The *Lutzomyia longipalpis* complex: reproduction distribution. In: Biosystematics of heamatophagous insect (eds: Service, M. W.). Clarendon, Oxford, pp. 257-269.

Wasserberg, G., Yarom, I. & Warburg, A. (2003). Seasonal abundance patterns of the sandfly *Phlebotomus papatasi* in climatically distinct foci of cutaneous leishmaniasis in Israeli deserts. *Journal of Medical Entomology,* **17**: 452–456.

White, G. B. (1982). Malaria Vector Ecology and Genetics. *British Medical Bulletin,* **38 (2)**:207-212.

White, G. B. & kellick-Kendrick, R. (1976). Polytene chromosomes of the sandfly. *Lutzomyia longipalpis* and the cytogenetics of Psychodidae in relation to other Diptera. *Journal of Entomology,* **50**: 187-196.

WHO, (1990). Control of the leishmaniasis. Technical Report Series 793, World Health Organization, Geneva, Switzerland.

WHO, (1998). Re- Orientation and definition of the role of malaria vector- control in Ethiopia. WHO. Geneva, Swizerland.

WHO, (2000). Division of control of tropical diseases. Leishmaniasis control home page www.who.int/health-topics/leishmaniasis.htm (updated 2000).

WHO, (2002). Annex 3: Burden of disease in DALYs by cause, sex and mortality stratum in WHO regions, estimates for 2001. In: *The World Health Report. Geneva: WHO, 192-197.*

WHO, (2006). Control of leishmaniasis- *Executive Board 11 8th Session.*

Wilkerson, R. C., Parsons, T. J. & Albright, D. G. (1993). Random Amplified Polymorphic RAPD markers readily distinguish cryptic mosquito species (Diptera: Culicidea: *Anopheles*). *Insect Molecular Biology*, **1(4)**: 205-311.

Wilkerson, R. C., Parson, T. J., Klein, T. A., Gaffigian, T. V., Bergo, E. & Consolin, J. (1995). Diagnosis by random amplified polymorphic DNA polymerase chain reaction of four cryptic species related to *Anopheles* (*Nyssorhymchus*) *albitarsis* (Diptera: Psychididae). From Paraguay, Argentina and Brazil. *Journal of Medical Entomology*, **32**: 697-704.

Williams, J. G. K., Kubelik, A. R., Rafalaski, K. J. A. & Tingey, S. V. (1990). DNA polymorphism amplified by arbitraty primers are useful as genetic markers. *Nucleic Acids Research,* **18**:6531-6535.

Williams, J. G. K., Reiter, R. S., Young, R. M. & Scolnik, P. A. (1993). Genetic mapping of mutations using phenotypic pools and mapped RAPD markers. *Nucleic Acids Research,* **21(11)**:2697-702.

Wright, S. (1943). Isolation by distance. *Genetics*, **28**:114-138.

Yang , Z. (1995). Evolution of several methods for estimating phylogenetic trees when substitution rates differ over nucleotide sites. *Journal of Molecular Evolution,* **40**: 689-697.

Yin, H., Mutebi, P., Marriot, S. & Lanzaro, G. C. (1999). Metaphase karyotyping and G-banding in sandflies of the *Lutzomyia longipalpis*. *Medical & Veterinary Entomology*, **13**: 72-77.

Zijlstra, E. E., & El-Hassan, A. M. (2001b). Leishmaniasis in Sudan. Post kala-azar dermal leishmaniasis. *Transactions of the Royal Society of Tropical Medicine & Hygiene*, **95 (Suppl 1)**: S59-76.

Zijlstra, E.E., & El-Hassan, A. M. (2001a). Leishmaniasis in Sudan. visceral leishmaniasis. *Transactions of the Royal Society of Tropical Medicine & Hygiene*, **95 (Suppl 1)**: S27-58.

Zijlstra. E. E., Musa, A. M., Khalil, E. A. G., El-Hassan I. M. & El-Hassan A. M. (2003). Post-kala-azar dermal leishmaniasis. *The Lancet Infectious Diseases*, **3**:87-98.

Appendix 1: RAPD primers used to screen *Phlebotomus orientalis* **form Sudan**

NO	Primer name	The sequence
1	Rap4b	TGACGATGCA
2	Rap13.2	ATTGCGTCCA
3	R38	CTAGCCGAC
4	R239	GCCTTTCCAG
5	R257.2	TGGAGCGGAA
6	R268	AGCTGGCTTG
7	R317.1	CGCTGAGATG
8	R356.4	GATCCCCTGA
9	R362	CGATCGACAC
10	R406.1	CGACATCCGT
11	R414	TCAAGCGATG
12	R455	GAGCCCTCAA
13	R469	TCGCAACGTC
14	R471.1	ACGGGTTAGT
15	R488.1	TGGACGGCAA
16	R512	TACACGACCC
17	R516.2	GAGTTCGCCC
18	R553	TAGAGAAGCC
19	R564.1	GCCTCCTACT
20	R727	TGTAGGTCCA
21	R739.1	GCTTACGAGG
22	R748	GCCGGTTTGG
23	R751.2	GGGCACTCCG
24	R769.2	CCTGGTCTAC
25	R782.1	CGAGTCAACT
26	R800.1	CTGGACCTTG
27	R807.1	GCCTTCATCT
28	R1059	CGTCGCTATT
29	R1302.1	GGAAATCGTG
30	R1305	CAGTGCGAAG

yes

I **want** morebooks!

Buy your books fast and straightforward online - at one of the world's fastest growing online book stores! Environmentally sound due to Print-on-Demand technologies.

Buy your books online at
www.get-morebooks.com

Kaufen Sie Ihre Bücher schnell und unkompliziert online – auf einer der am schnellsten wachsenden Buchhandelsplattformen weltweit!
Dank Print-On-Demand umwelt- und ressourcenschonend produziert.

Bücher schneller online kaufen
www.morebooks.de

OmniScriptum Marketing DEU GmbH
Heinrich-Böcking-Str. 6-8
D - 66121 Saarbrücken
Telefax: +49 681 93 81 567-9

info@omniscriptum.com
www.omniscriptum.com

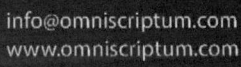

MIX
Papier aus verantwortungsvollen Quellen
Paper from responsible sources
FSC® C105338

Printed by Books on Demand GmbH, Norderstedt / Germany